D0929721

The Algorithmic Process

AN INTRODUCTION TO PROBLEM SOLVING

Gregory F. Wetzel
THE UNIVERSITY OF KANSAS

William G. Bulgren
THE UNIVERSITY OF KANSAS

SCIENCE RESEARCH ASSOCIATES, INC.
Chicago, Henley-on-Thames, Sydney, Toronto
A Subsidiary of IBM

The SRA Computer Science Series

William A. Barrett and John D. Couch, *Compiler Construction: Theory and Practice*
Marilyn Bohl and Arline Walter, *Introduction to PL/1 Programming and PL/C*
Mark Elson, *Concepts of Programming Languages*
Mark Elson, *Data Structures*
Peter Freeman, *Software Systems Principles: A Survey*
Philip Gilbert, *Software Design and Development*
A. N. Habermann, *Introduction to Operating System Design*
Harry Katzan, Jr., *Computer Systems Organization and Programming*
Henry Ledgard and Michael Marcotty, *The Programming Language Landscape*
Stephen M. Pizer, *Numerical Computing and Mathematical Analysis*
Harold S. Stone, *Introduction to Computer Architecture, Second Edition*
Gregory F. Wetzel and William G. Bulgren, *The Algorithmic Process: An Introduction to Problem Solving*

Acknowledgement *pp. 18–19*: George Polya, *How to Solve It: A New Aspect of Mathematical Method*. Copyright 1945, © renewed 1973 by Princeton University Press. Second edition copyright © 1957 by G. Polya. Excerpt adapted and reprinted with permission of Princeton University Press.

Acquisition Editor *Alan W. Lowe*
Project Editor *Molly Gardiner*
Designer *Carol Harris*
Technical Art *Ben DeBolt*
Composition *Graphic Typesetting Service*

Library of Congress Cataloging in Publication Data

Wetzel, Gregory F., 1950-
The algorithmic process.

Bibliography:
Includes index.
1. Electronic digital computers—Programming.
2. Algorithms. 3. Problem solving. I. Bulgren, William G.
II. Title.
QA76.6.W49 1985 001.64'2 84-23493
ISBN 0-574-21735-5

Printed in the United States of America.

10 9 8 7 6 5 4 3 2

Contents

Preface

> And though it is difficult to prescribe any Thing in these Sorts of Cases, and every Person's own Genius ought to be his Guide in these Operations; yet I will endeavour to shew the Way to Learners.
>
> Newton: *Universal Arithmetick*, translated by Ralphson, p. 198

The art and study of problem solving has lately been much neglected. However, its importance in today's highly technical world is widely recognized (though not always directly), despite its often conspicuous absence among the skills of students in colleges and universities around the country. Much of the blame for this gap in students' education is the tendency toward "cookbook" styles of presentation, especially in mathematics, chemistry, physics, and other topics in scientific and technical disciplines. Blame must also be apportioned to the humanities, whenever teachers fail to demand critical thinking and exposition of their students.

The result is that students no longer need to exercise their ingenuity in solving problems, but only to memorize certain forms and solutions. Missing in this type of learning is the need to abstract the essential qualities of a problem; to transform a problem by analogies and models to a more easily understood form; to question the premises of a problem; to see beyond the conventional limitations that might (incorrectly) be assumed to apply to a problem; to explore more than one possible solution; etc. Too often, in short, rote learning skirts the need to think, invent, explore, and discover.

This book will not attempt to teach students a "cookbook" method to solve problems. Indeed, it may be impossible to do so. Instead, it will attempt to get the student to exercise his or her abilities by posing problems for solution, by suggesting aids in solving problems, by demonstrating how the authors would solve a problem, and by describing a method that allows a solution to be rigorously and unambiguously written out.

Our primary purpose is to provide an introduction to problem solving techniques that are particularly useful in using a computer in a problem's solution. This has been called *computer programming*. Most people, however, do not realize that the human problem solver has the lion's share of such work. To attempt to teach programming by teaching a person how to "talk to" or "interface with" the machine ignores the fact that men and women may need to revise or expand their own problem solving skills substantially before they attempt to use the machine. Toward this end, we treat only one of the two classes of formal problems: problems of analysis, not problems of synthesis.

A further difficulty that computer science teachers encounter in students is near total ignorance of the history of machines and of the history of computing. This is no one's fault, but implies that students are not culturally prepared for some of the important ideas behind writing computer programs. It seems to the student that some of the ideas are too foreign or "high brow" to assimilate, and it seems to the teacher that the student is slow. Neither is true; it is merely a matter of acculturation. Because most of these concepts are neither new nor familiar, we attempt to provide a background through historical anecdotes (e.g., the Jacquard loom). By doing this, we hope to encourage students to adopt the attitude that these concepts have been around for years and so are accessible to anyone, not just PhDs who read the latest journals.

This book can be used in an introductory computer science course, either as a stand-alone text, in which no programming is to be done by the student; or in conjunction with a text in which programming projects must be done. In either case, we believe, this book approaches the real problem that people face: not a solution in a computer language; but solving a problem in a way that's *amenable* to expression in a computer language.

No mathematical background is presumed in order to understand the material in this book, except high school algebra. We suggest, however, that the skills of forming and manipulating abstractions and in using analogy, which are stressed in mathematics classes, are important in understanding the material presented here. Students who lack such background should be sure they understand the examples throughout the text, before proceeding further, and to build on that understanding by doing as many exercises as possible.

Chapters 1 through 5 should be covered before a particular programming language is discussed—if programming projects are to be given.

In our course at the University of Kansas, we teach the programming language Pascal, and use a second textbook to cover that material. We cover chapters 1 through 5 in two to three weeks (6 to 9 hours of lecture). Then, as we introduce the programming language, we constantly refer back to the discussion on algorithms that has gone before. Thus we attempt to accomplish two goals: (1) stress the important material—the understanding of algorithm development—twice, once by itself and again in terms of a programming language; and (2) separate the two distinct topics, algorithm development and programming language syntax, to avoid confusion.

We find that the latter goal is particularly important. Our students find it much easier to learn the two topics if we don't try to teach programming language syntax at the same time we try to teach algorithm development. We cover the basics of program syntax (excluding arrays, records, procedures, or functions) and the mechanism for running and debugging programs on a computer in another two or three weeks.

Chapter 6, Fundamental Data Types, is mainly for those who use this book without programming exercises. If a programming language is taught, we find it more useful to use the supporting text, though it sometimes helps students to see the material presented in more than one way.

Chapters 7, 8, and 9 introduce subalgorithms and structural abstraction (top-down design and modular refinement), aggregate data types (arrays and records), and prototype algorithms for searching, sorting, etc. (using arrays and records). As before, we introduce these topics on their own, using the material in this book, and again when we describe them in terms of a programming language (alternating between the two). This section of our course takes six to seven weeks (18 to 21 lectures).

Chapter 10 and appendix A cover the efficiency and analysis of algorithms and program stylistics. Appendix A can be covered at any point after chapter 2.

Appendices B and C provide translations of the major algorithms developed in this book. Appendix B provides translations into Pascal and appendix C provides translations into FORTRAN 77.

The authors and publisher would like to thank the following reviewers for their valuable feedback:

Professor Charles C. Weems, *University of Massachusetts*
Michael Marcotty, *General Motors Research Laboratories*
Dr. Bill McBride, *Baylor University*

1
Problems and Problem Solutions

> The solution of problems is the most characteristic and peculiar sort of voluntary thinking.
> William James

1.1. WHAT IS A PROBLEM?

Many times we face situations in which we want or need to obtain some goal or the means to some end. If we quickly and easily achieve the goal or find the means, we do not have a problem. We have a problem only when the goal is not immediately achievable or the means are not obvious or easily found.

People deal with problems everyday: car problems, health problems, social problems, ethical problems, and (not the least) money problems. These sorts of problems we shall refer to as **informal problems** (which is not to say that we solve them while wearing blue jeans or in a casual way). By *informal* we mean that we do not necessarily find solutions by precisely specifying the initial conditions, the desired results, or the actions by which we achieve the desired results.

In this book, we are interested only in **formal problems**, because these are the kinds we can solve by writing a program for a computer. In contrast to informal problems, formal problems are characterized by complete specifications: precisely specified initial conditions, as well as solutions or results of a specified form; and they must be solved by a completely specified set of actions. We draw a distinction between the *procedure* used to solve a problem and the *actual results* obtained by using the procedure to solve a *specific instance* of the problem. Given this description, you might guess that artificial problems, such as games, are really formal problems.

Consider, for example, the game of Monopoly. The game is a problem because you desire to win, but you don't know how to win with certainty. It is a formal problem because every detail is specified. The initial conditions are of a starting position on the board and a specific amount of Monopoly money for each player. The desired result is to win by forcing everyone else into bankruptcy, and the rules of the game form the complete set of permissible actions. You, the problem solver, must solve the problem (win the game) by acting according to the rules.

That is, at the time specified in the rules (your turn), you choose to perform one or more of the actions allowed by the rules under the conditions of the game at that time. With luck in rolling the dice and wisdom in your choices of action, you will solve the problem (win the game). We assume, of course, that you win because you make the best choices ingeniously and not because you stumbled into winning it without understanding.

Formal problems can be classified into two categories: problems of synthesis and problems of analysis. Problems of **synthesis** have specific initial conditions and specific plans of actions, but specify only the general form of the result. (After all, something must be unknown or we wouldn't have a problem.) Examples of this type of formal problem are easy to find in mathematics, and are referred to as "to find" or "to prove" problems.

An example of a "to find" problem is the system of two simultaneous equations:

$$x = 3 - y$$
$$y = x + 7$$

The initial condition is the statement of the two equations. The specific plans of action are the two common ways of solving such systems of equations (substitution or cancellation by subtraction). The form of the solution is:

$$y = \text{(some unknown)}$$
$$x = \text{(some unknown)}$$

In this problem, we apply the rules of algebraic manipulation to the initial conditions in the sequence described in an algebra textbook, and out pops the result:

$$y = 5, x = -2.$$

Problems of **analysis** are those in which we know the initial conditions and results, but not a specific plan of action. That is, we don't know how to get from the initial conditions to the results, or we don't know what actions we may use. The problem solver must set down a plan of action, choosing only from the specified set of allowable actions. (This sequence of actions will later form the plan for synthesis.) These sorts of problems are usually solved by working backward from the result to the initial conditions. We discern differences between the initial conditions and the results and, by taking one action at a time, attempt to reduce the differences.

An example of this sort of problem occurs when you're driving in an unfamiliar city. You know where you are (the initial condition) and you know where you want to be (the result), but you don't know how to get there. In this example, the set of allowable or permissible actions is: drive straight ahead some fixed distance, turn right, or turn left. If you were to ask someone for help to solve your problem, he or she might respond by giving you a plan or sequence of actions that, if executed, would lead you to your destination. On the other hand, the person might reply, "I don't know; I'm lost too!" In this case, you both have a problem.

EXAMPLE 1.1. Finding Your Way in a City

As an example, let's consider the problem of finding your way around an unfamiliar city. Suppose you know these facts about the city you're in:

1. Numbered streets run east and west.
2. Numbered streets have larger numbers as you go south.
3. Named streets run north and south.
4. Main Street is a major thoroughfare, running the entire length of the city.
5. Addresses on numbered streets are given by the number of blocks east or west from Main Street, plus the house number (e.g., 659 W. 10th St. is 6 blocks west of Main on 10th Street).
6. Streets east of Main are named after things or people. Streets west of Main are named after places.
7. You are at the corner of 19th St. and Iowa St.
8. You need to go to 531 Elm.

To solve this problem, we could reason this way. Block 5 is north of block 19. Since we are at block 19, we need to go north. Elm is the name of a thing, so it is east of Main St. We are on Iowa St., so we must be west of Main St.; therefore we need to go east. Since we know that Main St. goes all the way through the city, we should take it north (so we won't hit a dead end). Our plan, then, is the following:

1. Take 19th St. east to Main
2. Take Main St. north to 6th St
3. Take 6th St. east to Elm St
4. Take Elm St. north to 531 Elm

Of course, this presumes several things: that you know which direction is north; that 19th St. is open (or continuous) from Iowa to Main; that 6th St. is open from Main to Elm; and that Elm is open at 6th St. However, we should always be aware that things will go wrong, and plan accordingly. Furthermore, our plan, even if it works, is not stated by the permissible actions given above (turn right, turn left, etc.). So we need to restate our solution by using these actions, even though we solved our problem by using other actions. Since our actions are more general than the permissible actions, we have to refine our solution, in addition to restating it. Once we have done this analysis, the last step, synthesis, is merely to drive our car according to our plan (solution).

All problems that are solved with the aid of a computer are problems of synthesis. Most problems solved by a computer *programmer* are problems of analysis. A programmer's task is to transform a problem from one of analysis to one of synthesis by specifying a general plan of action. Computer programmers know what kinds of initial conditions (data or inputs) they should expect and they know the result for different sets of data. They then write the program—that is, a plan or sequence of actions that, when executed by the computer, starts with the initial conditions and arrives at the result. Since we are primarily interested in problem solving as it applies to computer programmers, we will concentrate on formal analytic problems and ignore the rest.

1.2. WHAT IS A PROBLEM SOLUTION?

> After the Learner has been some Time exercised in managing and transforming Equations, Order requires that he should try his skill in bringing Questions to an Equation.
> Newton: *Universal Arithmetick*, translated by Ralphson, p. 174

Given that we are interested only in formal problems of analysis, what, exactly, constitutes a solution? To answer that question, we must get a better grasp on what, exactly, is a problem of analysis. The Greek mathematician Pappus (who lived around 300 A.D.) discussed the distinction between analysis and synthesis, and below is a paraphrase (by G. Polya from Pappus' "Analyomenos," which means "Treasury of Analysis").

> In analysis, we start from what is required, we take it for granted, and we draw consequences, till we reach a point that we can use as a starting point in synthesis. For in analysis, we assume what is required to be done as already done (what is sought as already found, what we have to prove as true). We inquire from what antecedent the desired result could be derived; then we inquire again what could be the antecedent of that antecedent, and so on, until passing from antecedent to antecedent, we come eventually upon something already known or admittedly true. This procedure we call analysis, or solution backwards, or regressive reasoning.

> But in synthesis, reversing the process, we start from the point which we reached last of all in the analysis, from the thing already known or admittedly true. We derive from it what preceded it in the analysis, and go on making derivations until, retracing our steps, we finally succeed in arriving at what is required. This procedure is called synthesis, or constructive solution, or progressive reasoning (G. Polya, *How to Solve It*, p. 142).

From this we see that the *method* of solving the problem is the *solution* to our problem. In this respect we have a *metaproblem*—that is, a problem that concerns the aspects of other problems. In computer programming, we usually are not interested in the specific steps for solving a specific problem. Such a problem is usually more easily done by hand (than by computer). Problems that would take too long to do by hand—that require that observations be taken faster than any person can perform them, etc.—are the exceptions. Rather, we are interested in specifying a method, a plan, or a procedure to solve a family or class of problems.

EXAMPLE 1.2. Summing a List

Find the sum of any list of numbers—say 3, 5, 1, 2. Of course, you will say that this is hardly a problem at all; the answer is 11. But you must remember that we are not interested in the sum, 11, but in the sequence of actions by which you find the sum, given the numbers. Think about it, and write the steps down before you read further.

You probably wrote something like this:

Add the 3 and the 5 together, getting 8.
Add the 8 to the 1, getting 9.
Add the 9 to the 2, getting 11 (the answer).

This is fine, except that it is not very general (and it's not the right answer). This sequence of actions will work only for the list 3, 5, 1, 2, or any reordering of that list (the order is unimportant here). Eleven is not the right answer because the steps do not solve the problem. They solve the problem of adding 3, 5, 1, and 2, but not the problem of finding the sum of *any* list of numbers.

A more general approach would be to specify the actions like this:

Add the first and second items together, getting a partial sum.
Add that partial sum to the third item, getting a second partial sum.
Add the second partial sum to the fourth item, getting the final sum.

This also is fine, but is *still* not the right answer.

The problem does not specify that the list is exactly four items long. It might be three items long; it might be 1000 items long; it might be zero items long! In none of these cases would our suggested procedure produce a correct sum. The version above will find the sum of any *four* numbers, but not the sum of *any* list of numbers.

We must be careful not to make assumptions about a general problem simply because we make them about an example or instance from a family of problems we are trying to solve. We are looking for a general procedure to solve any problem that involves summing a list of numbers.

This discussion of problem solutions should provide the "flavor" of what's to come. Before we can talk seriously about problem solutions, we must agree upon a set of actions that constitutes the "permissible" or "allowable" actions referred to in the previous section. Without knowing what actions we are allowed in our procedure, we will not be able to specify a proper sequence of actions to solve the problem. (We will return to our discussion of problem solutions in a later chapter.)

1.3. PROBLEM SOLVING AND THE COMPUTER

People have been using machines to help solve problems for thousands of years. The twist we have introduced in modern times is having a machine follow a set of external directions to accomplish a task. In the old way, a machine is constructed to perform a specific sequence of internal actions to solve a specific problem. Thus the hands of a clock go around to track time and a wagon's wheels allow it roll. They can do only the one thing they are constructed to do. Now, however, we construct machines that have a repertoire of actions. From this repertoire we select a sequence of actions to suit our purposes and instruct the machine to follow them.

This is an abstract way of saying something we do nearly every day. For example, we construct cars to perform certain actions, and their repertoire of actions is to move forward or backward, at varying speeds; to turn left or right, to varying degrees; to honk the horn; signal turns; light the way in front; etc. While driving, we select the action we want at any moment by depressing the accelerator or brake, selecting a gear, turning the steering wheel, pressing a button, flipping a switch, etc.

But this is not the height of our ingenuity. We have constructed machines that can accept a list (sequence) of actions and perform them, one after the other, without a person to push the right button at the right time, turn the wheel, or whatever. One of the first such machines, the Jacquard loom to weave fabric (1805), could be given a sequence of pasteboard cards in which punched holes directed the mechanism within the loom. In a sense, it was the first robot, and is still in use today (in an updated form). To create a patterned weave, a weaver (the analog of today's programmer) determined how to punch holes in the cards.

The cards were then tied together and the first card was placed in the loom's "reader." From then on, the machine did the rest.

The advantages of such machines are the same today as in 1805. Once the weaver punched the "program" on the cards, the loom could weave that pattern over and over again, without need of a highly skilled person. It also freed weavers from tedious hours over a loom, allowing them to spend their time more constructively. The machine's time was used more effectively, too. It could produce cloth much faster and with fewer flaws. Furthermore, since weavers were freed from manually producing cloth, they could design, and have machines execute, patterns more complex than weavers could handle before.

The most sophisticated machines today are computers. Like the Jacquard loom, the computer can accept a list of instructions. There are important differences, however, the least of which is that a computer is not a loom.

The greatest difference is the "memory" of a computer. It is impossible for a loom to "remember" where it was or what it was doing, or to leave a note for itself to "look" at later. In other words, no action could depend upon a value or condition determined by a previous action. But a computer has memory cells, each of which holds a single value. (A value may be a number, a character, or true or false.) Actions in the computer's repertoire store values into memory cells and retrieve copies of those values from the memory cells. Without this kind of memory, we would be unable to write programs to solve more than simple problems.

Memory is particularly useful because of this second major difference: a repertoire of actions, which includes **conditional actions** (sometimes called **conditional instructions**). This means conditional instructions (actions) compare the values of two specified memory cells. Based on the comparison, conditional instructions direct the computer to perform one of two or more groups of instructions.

These two additions allow us to write programs and algorithms (specify sequences of actions) to instruct a computer to do almost anything. We must be ingenious enough, however, to devise the programs. Thus computers are limited by the talents of their human programmers. Some people contend there are problems that humans can solve that computers inherently can't. These are usually of a metaphysical or problematical nature.

At this point, we should begin to distinguish among three words: *plan*, *algorithm*, and *program*. In subsequent chapters, these words will recur many times, and we should recognize their meanings.

A **plan of action** is a general outline of the major steps we need to take to accomplish our goal, starting at a predefined initial point.

An **algorithm** is a refinement of a plan—a statement in finer detail. In an algorithm, we may use only well-defined steps that do not allow for uncertainty. All choices are made by using explicitly stated criteria; no "maybe's" are allowed.

A **program** is the statement of an algorithm in a computer programming language (C, Pascal, FORTRAN, etc.). The computer cannot accept most plans or algorithms and act on their instructions. Whereas a plan might not be sufficiently specific or explicit, an algorithm might not use the right "words" or punctuation to direct the computer's circuitry. A computer can only process a program that is written in the rigid syntax of a programming language.

This book provides an introduction to problem solving and algorithms, not to any programming language. We have found, however, that writing plans and algorithms is indispensable for writing good programs. The idea is to worry about solving a problem, and not state it in so much detail that you lose your train of thought.

1.4. SUMMARY

In this chapter we have classified problems, according to their characteristics, into formal and informal. This classification is based on our ability to specify initial conditions, results, and permissible actions.

We subdivided the class of formal problems into two sub-categories: synthetic and analytic. To solve a synthetic problem, we need to know the initial conditions and a plan of action that would transform them into the answer. To solve an analytic problem, we need to know the initial conditions and the form of the result or answer. Then we have to devise a plan to transform the initial conditions into the result, and carry out the plan to solve a synthetic problem.

We also discussed problem solutions as a general plan of action. We need to transform analytic problems into synthetic problems by specifying a plan so that a machine, the computer, can finish the solution by following a plan. We discussed the need for generality in the plan of actions.

Finally, we discussed machines and their instruction repertoires. We also introduced the concepts of memory and conditional instructions and the distinctions between plans, algorithms, and programs.

1.5. EXERCISES

1.01. Indicate whether each problem below is a formal or an informal problem. If a formal problem, give the initial conditions, the form of the results, and a set of permissible actions. (Do not try to solve the problem; simply classify it.)

a. You have been given a recipe, but whoever wrote it forgot to say what it makes. What does it make?

b. You have applied for a scholarship but have not been notified of the result. The scholarship would pay for all expenses, but today is the last day to apply for loans for next year. If you apply for a loan, you automatically waive the scholarship. What should you do?

c. You have bet an acquaintance $20 that "your team" will win tonight's game. Your acquaintance is 6 feet, 5 inches tall and weighs 225 pounds (the Packers are interested in him). You do not have $20. What should you do?

d. You need the cube root of 19384.3 immediately, but left your calculator at home. What is the cube root of 19384.3?

e. You are given a map of Kansas and told to find the shortest route from Lawrence, Kansas, to Liberal, Kansas. The map is marked with mileages between junctions. What is the shortest route?

1.02. All the problems below are formal problems. Indicate whether they are synthetic or analytic problems. (Do not try to solve the problems; merely classify them.)

a. Your bank has several electronic tellers, and you are using one for the first time. Your checking account balance is $100, and you need to withdraw $20 (see part c of the previous problem). What do you do?

b. An accomplished pianist, who has been given a newly composed piece of music, is asked if he likes it. How does he respond?

c. It is Saturday afternoon and you're watching a college football game. There is one second left in the game, and the team with the ball is 5 points behind, on its 20-yard line. What should the coach do?

d. A new night watchman on his rounds, approaches a locked door. He wants to unlock the door but there must be a million keys on his key ring. How does he finish his rounds on time?

e. Sherlock Holmes, upon completing a case, met with Dr. Watson, presented the essential facts to the good doctor, and suggested that he attempt to deduce the villain in the case. By using the rules of deduction so well defined by Holmes, Watson succeeded. Who was the villain?

1.03. You work for ACME Merchandising and are about to run a sale: 13% off everything. (It's the store's 13th birthday and the manager, who's not superstitious, wants to celebrate.) Unfortunately, the clerks have a hard time figuring 13% in their heads. You receive a list of prices from each department and must write a procedure (carried out by your faithful assistant) to compute 13% off each price. The result is a list of the original prices and the new prices (after discount), which you are to send to the departments. Write the procedure you will give to your assistant.

2

Algorithms, Problem Solving, and Heuristics

Then let him give Names to both known and unknown Quantities, as far as Occassion requires, and express the Sense of the Question in the Analytick Language.

Newton: *Universal Arithmetick*, translated by Ralphson, p. 174

Solving problems is a fundamental human activity. In fact, the greater part of our conscious thinking is concerned with problems. When we do not indulge in mere musing or daydreaming, our thoughts are directed toward some end; we seek means, we seek to solve a problem.

G. Polya: *How to Solve It*, p. 221

This chapter provides a further discussion of problem solving and, in particular, the use of heuristics in problem solving. **Heuristic** derives from the Greek word *heuriskein*, "to find." A heuristic is a rule or question that often leads to fruitful lines of thought.

The types of problems we will attempt to solve are ones amenable to computer solution; since we're just beginning, the problems we will face are not hard to solve. We suggest no problem that takes any real time to do by hand. We are more interested in analyzing solutions, and expressing them rigorously and unambiguously, than in the answers.

All of this, of course, is leading to writing computer programs—which, unfortunately, is not easy. Its difficulties are in discovering how to start from some initial point, with a given set of data, and to obtain a final result. We will concentrate on formal analytic problems.

2.1. AN INTUITIVE DEFINITION

Most people encounter algorithms or something that resembles an algorithm every day. Simply put, an algorithm tells us how to solve a certain type of problem, and each of us probably solves problems every day by following such a set of rules or directions.

For example, when we purchase a candy bar from a vending machine, the directions might say to deposit 40 cents in a certain slot, then press a selector button (indicating a certain candy) and the candy bar will appear in a designated slot. Another set of directions applies if the plan fails to produce a candy bar.

Later in the day, returning home, we decide to prepare dinner, and the recipe we choose may appear as follows:

Mom's Guaranteed Chili

Brown 1 lb. ground beef
Add one medium onion, chopped
Add one 15 oz. can tomatoes & one 15 oz. can kidney beans
Season with 1 tsp. salt and 1 T. chili powder
Simmer 1 to 1-1/2 hours
Serve with crackers

ALGORITHM EXAMPLE 2–1

Most people successfully follow such directions because they are easily understood. However, training is sometimes required before directions can be understood and executed, so that one must be trained in a special language or must know special abbreviations. For example, consider the directions below.

Standardization of a Solution of NaOH

1. Pipette 20 ml of unknown acid into a clean, dry flask.
2. Titrate, using a standardized solution of NaOH (approximately 0.1 N), to phenolpthalein end point.

3. Record total volume NaOH used.
4. Compute milli-equivalents of NaOH used by the formula:
   meq NaOH = (ml NaOH used) × (Normality of NaOH)
5. Assuming the unknown is diprotic, compute the amount of unknown acid present by the formula:
   millimoles acid = (meq NaOH) / 2

ALGORITHM EXAMPLE 2–2

Many find this set of directions meaningless, but if one is trained in the language and abbreviations common in chemistry, the directions are easy to follow, and one can determine the amount of acid in a sample of unknown concentration. In other words, directions for performing an experiment in chemistry form a type of algorithm: a set of instructions we can use.

We learn how to solve problems of many types, provided we understand the language and rules associated with a particular skill, and this applies especially to computer programming. A person who uses a computer must learn a special language. After training, the student can write a precise description of a solution to a class of problems—that is, construct an algorithm.

Some of these algorithms and sets of rules are simple; others are complex. With training, however, we recognize the similarities among a class of problems. Even though problems may seem complex, we can develop a general method or common algorithm for solving a class of problems if we understand their similarities.

2.2. AN EXAMPLE: FOLLOWING A PLAN

When we use the computer to implement problem solutions, we find that two features are especially useful in developing directions or algorithms. These features are the decision-making and arithmetic capabilities of the computer.

In an example that employs these important attributes of the computer, we will use a standard algorithm that specifies that the computer should perform an initial instruction. Instructions will then be performed sequentially, except when we need to do one sequence if a condition is true and nothing, or some second sequence, if the condition is false. Also, the arithmetic functions of the computer will calculate the values needed to arrive at a solution.

Let us now turn to the algorithm and follow it through "by hand," mimicking what the computer does. (The algorithm specifies how the computer's decision-making and arithmetic capabilities are to be used.) This example also illustrates how an algorithm is written. You need no knowledge of the problem to execute the algorithm, only experience in the use of a calculator. In addition to the algorithm, you have a list of data values called the **input list**.

1. Read a number from the input list, and let *x* be the name of a memory cell whose initial value is this number.
2. Write the value in memory cell *x* on a clean sheet of paper. This sheet is called the **output answer sheet**.
3. If the value in *x* is less than zero, execute instruction 11. Otherwise, continue to instruction 4.
4. Let *s* be the name of a memory cell whose initial value is 1.
5. Let *i* be the name of a memory cell whose initial value is 3.
6. Calculate the value of $(s + (x/s)) / 2$. Let *t* be the name of a memory cell whose value is the calculated quantity; that is, set *t* equal to the value of $(s + (x/s)) / 2$.
7. Assign the value in *t* to memory cell *s*. That is, set *s* equal to the value in *t* and forget the *old* value associated with *s*.
8. Decrease the value in memory cell *i* by 1.
9. If the value in *i* is positive, execute instruction 6. Otherwise, continue to instruction 10.
10. Write the value in *s* on the output answer sheet.
11. Stop; you have finished

ALGORITHM EXAMPLE 2–3

Doing the calculations with pencil and paper, follow the instructions in the example, for some initial value. If the initial value is 2, for example, the values of the memory cells after the first five instructions have been performed are x contains 2, s contains 1, i contains 3. Thus the result of instruction 6 is t contains 1.5. After instruction 8, we find x contains 2, s contains 1.5, i contains 2. Notice how the initial value of i determines the number of times instructions 6 through 9 are performed.

In the next two times through this *repetition* of four instructions, the value in s is changed to 1 and 5/12 and then to 1 and 169/408. The last value in s is 1.4142 (rounded to four decimal places). You may recognize this number as approximately the square root of 2.

This list of instructions is a simplified version of a commonly used procedure for calculating the square root of a number. The general procedure is called the *Newton-Raphson* algorithm for calculating square roots. The important point is that it is possible to follow the set of instructions and, hence, calculate an approximation to the square root of a number without any knowledge of the underlying theory, of what is accomplished by following the set of instructions, or even of what the number might be. It is precisely this sort of obedience to instructions, and *only* this sort of obedience, that you can expect from a computer.

2.3. ALGORITHMS AND HUMANS

As you may have surmised from the example algorithm for extracting a square root, algorithms are not a natural (or at least usual) way of stating a problem's solution, because we usually do not state our plan of action. Rather, we execute it as we think of it, which leads to the following.

First, we are used to tailoring our plans of action to the particular problem at hand, not to a general problem (of which the particular problem is an instance). This results in "nearsightedness" in problem solving: we are not used to abstracting or generalizing our problem statements. Thus we miss many salient details that would lead to a better understanding and, hence, to a better solution of problems. (Often, a statement of a generalized problem is easier to comprehend than a specific problem, and sometimes easier to solve.)

Second, because we usually do not write out plans of actions, we do not use the science that has been developed for solving problems. Because we are unaware of the basic ideas we use to formulate plans of action, we hardly think about them: we just "do it."

This means that to write algorithms that are comprehensive, precise, and accurate, we need to relearn old ways of doing things and to discard—perhaps even realize that we *have*— old habits. Since an algorithm is self-contained, we must become aware of all the assumptions we make, in addition to the initial conditions. We must be careful not to overlook steps in a procedure just because they "make sense." Machines don't have common sense to apply to problems.

2.4. A FORMAL DEFINITION

We should be rigorous when we discuss algorithms. To this point, we have been informal about the specifics of algorithms; our aim has been to build an intuition about algorithms. There is, however, a formal definition for an algorithm and there are formal ways of defining problems.

An **algorithm** is a finite sequence of unambiguous instructions that, when given a set of input values, produces the desired outputs, then stops.

This definition is complete, but many important properties are only implied, rather than explicit. What does it mean to say that an algorithm is a "finite sequence"—that an algorithm's statement must not go on forever, that every algorithm has a *definite length*? What are the ramifications?

If an algorithm has a definite length, we can write it down, which would not be the case if it were infinitely long. More important, if an algorithm has a fixed length, but it can handle computations that are infinitely or arbitrarily long, we can deduce that there must be a mechanism for repeating instructions.

If this seems unlikely, consider the problem of counting the number of cars that pass your home in one hour. Since you don't know the number, you must conduct an experiment.

EXAMPLE 2.1. Counting Cars

Suppose you have a hand counter (like those used by ticket takers at museums and sporting events) and a stopwatch. We could write an algorithm to describe the experiment, as follows.

```
reset the stopwatch
set the counter to zero
start the stopwatch

wait until a car passes
if the stopwatch shows less than 1 hour
  then
    push the button to increment the counter
  else (if the stopwatch reads 1 hour or more)
    stop and report the value in the counter
end-if

wait until a car passes
if the stopwatch shows less than 1 hour
  then
    push the button to increment the counter
  else (if the stopwatch reads more than 1 hour)
    stop and report the value in the counter
end-if

wait until a car passes
    .
    .
    .
    etc.
```

ALGORITHM EXAMPLE 2–4

The trouble is, we don't know how many times to repeat the instructions (from "wait" to "stop"). If we decide to repeat them no more than 50 times, we will be wrong whenever there are *more* than 50 cars. We might be tricky and repeat the instructions a ridiculously large number of times, but that will work only for this problem and is a waste of effort.

The point is that if the list of instructions is finite and the number of cars is indefinitely large, there must be a way to repeat a group of instructions without writing it again and again. Later, we will see how to do this.

The meaning of: "an algorithm must eventually stop" seems clear, but what is implied by this part of the definition? We have seen that the length of an algorithm is finite, and (for the moment) let us ignore the possibility of repeated instructions, so that we have a sequence of instructions of a definite length with no repeats. Must such a sequence always stop?

You might reply that if the agent that executes an algorithm cannot go back to an instruction it has passed, it must inevitably come to the end. True enough; but only if each instruction can be performed in a finite amount of time—that is, if we don't get stuck at one instruction forever.

Can you think of a very simple procedure that will not stop? (Consider what happens if a car *never* passes—you will wait forever.) A sequence of instructions is not an algorithm if it does not stop with even one event—for one specific set of input values (in this case, the "empty set").

This same reasoning applies to repeated groups of instructions, called **loops**. So the algorithm can stop, each loop must eventually stop.

Lastly—and perhaps for our purposes most importantly—the definition demands that each instruction be unambiguous. This means that each instruction must perform a precise task. Less obviously, this also means that each instruction will do only what we tell it to do.

An instruction that only partly performs a task cannot be used in the hope that it will "magically" do it all. Since a computer is a machine that faithfully and unerringly does precisely what you tell it, you will succeed in programming it only if you know—not guess—exactly what you are saying and doing.

In what follows, we shall help you break old habits by encouraging new ones. To write good algorithms, you need to be inquisitive. It helps to ask tried-and-true questions, and we will suggest some. It also helps to see how old ways fail where new ones succeed; so we will work examples. The best way to learn this material is to use it. Work problems; apply this learning in other endeavors. It has been said that we never really learn anything, we just get used to it. Get used to it!

2.5. TOWARD A PROBLEM SOLUTION

Many people think that computer programming is mathematical, but it is not. In mathematics, given a set of conditions, the proof of a theorem exhibits the truth of its statement, subject to the conditions. In computer programming, given a set of initial data and the *form* of the result, we show a sequence of actions, or an algorithm, that if faithfully followed will transform the data into the required results. We also find in computer programming, that we must use only operations the computer can perform in expressing our problem solution. In this regard, computer programming is like playing a game, because the object is to win without breaking the rules. Actions must be limited to those that are allowed by the rules.

In solving any problem, no matter how simple, we must perform several steps. Although many people have attempted to outline rules for solving analytic problems, this has stumped the greatest intellects in history. If they had succeeded, *all* problems would be reduced to problems of synthesis. As it happens, we can only ask questions that may lead to insights that might not otherwise occur to us. Also, there are stages or steps in solving a problem, and most people agree on a general plan that involves four steps (modified from *How to Solve It*, by G. Polya).

First. *UNDERSTANDING THE PROBLEM.*

What is the required result? What are the data? What is the starting point, or the initial conditions? What are the rules? (What are the allowed operations?)

Is it possible to get the result from the data (with or without following the rules)? Do the problem by hand. What steps did you take? Can you write them down?

Is the problem divided into major parts? What are they? How do they fit together (what memory cells do they share)?

Have you assumed conditions about the problem that are not specified? What are they? Do they restrict the generality of the problem? Does this make a difference?

Second. *DEVISING A PLAN.*

Have you seen it before, or the same problem in a slightly different form?

Do you know a related or previously solved problem that could be useful?

Look at the result. Try to think of a familiar problem that has a similar result.

Here is a problem related to yours and solved before. Could you use it? Could you use part of it? Could you use its method? Can you modify it so you can use it?

Look at the data. Repetitions in data hint at loops in the solution. What are the patterns?

If you cannot solve the proposed problem, try to solve first some related problem. Can you imagine a more accessible related problem? A more general problem? A more special problem? An analogous problem? Can you solve part of the problem? Keep only part of the result and drop the rest: How much can you solve of the modified problem? Can you think of a new rule that would help to solve the problem? Can you restate the operation of the new rule in terms of rules you already have?

Third. *CARRYING OUT THE PLAN.*

Check each step. Can you see clearly that each step is correct? Have you considered special cases?

Can you check the result? Is the "answer" correct? Can you arrive at the right result, given a reasonable set of data? Given an unreasonable set of data?

Make certain your solution works for "boundary" conditions. Does it work if the data list is empty? Too long? Can you acquire data that results in division by zero (or other nonsense)?

Fourth. *LOOKING BACK.*

Do you compute an intermediate result that isn't used later? Can you eliminate it to simplify the solution?

Can you derive the result differently? Can you make your solution simpler or more general?

Can you use the solution, or method, for another problem?

2.6. USING HEURISTICS

Consider the problem of finding the average of a list of numbers—a simple problem, but it embodies many properties of problems in general. Because it is easily understood, there is no difficulty understanding what we are trying to do (even if at times you may not understand how we go about it). At first glance it seems easy. All you do is add them up, count them, and divide the sum by the number of items. As far as this goes, it's right. Unfortunately, however, we haven't followed the rules of the game. In other words, what operations are you *allowed* to specify? (*Hint:* "Add them up" is not one of the rules.) Computers, because they are rather simple machines, aren't capable of doing many things at one time. For this example, let's discuss **named memory cells**. One of the rules in using a computer is that you are allowed to add a number to a named memory cell (as in step 6 in Example 2–3). You can also store a number, or a character, in a named memory cell (as in steps 4, 5, 6, and 7). Other common computer operations are multiplication, division, addition, and subtraction. Two other important operations are reading a number or character off the data list, (storing the item into a named memory cell), and writing (or printing) the contents of a named memory cell.

Why keep using *named memory cell*? Because the computer is a machine with memory cells. You're outside the computer, and you can't see the cells, even if you open the computer. So if you can't point to the cell you want to use, how can you refer to it if you don't give it a name? This is a fundamental rule in computer usage: Everything has to have a name, and you must spell it the same way every time or the machine will think you mean something different.

Let's return to our problem of finding the average. Play by the rules this time and see how close you come to solving the problem.

EXAMPLE 2.2. Averaging a List of Numbers

One of several possible solutions goes like this. First we have a memory cell we call SUM. Second, we think of what we do to find the average of a list of numbers using only the allowable operations. (*Hint*: What keys would you hit on a calculator while finding the average?) The first thing you probably do is clear SUM to zero. (The necessity of this step will become obvious in a moment.) Third, get the next number on the data list, and fourth, add it to SUM. (If you don't start SUM out at zero, what are you adding the first number to?) Repeat these last two steps until you finish the data list. At this point, you have the total of all the numbers in SUM. (We have also described how to "add them up"—which is the way we wanted to say it initially, but couldn't.)

Finally, divide SUM by COUNT and write out that result. Where did COUNT come from? Moreover, how many numbers were there?

In a neater format, let's write out the algorithm we have thus far:

```
1. Set SUM to 0
2. Repeat steps 3 and 4
3.    Read the next number off the data list
4.    Add the number just read to SUM
```

ALGORITHM EXAMPLE 2–5

We have to count the numbers as we read them off the data list. To remember that count in the algorithm, we need another memory cell (call it COUNT). Now we must add to the algorithm so that the instructions arrange for each number to be counted as it is read. This involves setting the counter to zero and adding one whenever we read a number. So modify the algorithm to obtain:

```
1. Set SUM and COUNT to 0
2. Repeat steps 3-5
3.    Read the next number off the data list
4.    Add the number just read to SUM
5.    Add 1 to COUNT
```

ALGORITHM EXAMPLE 2–6

Now the last step, we finish the algorithm to find the average: adding the last step,

```
6. Compute the average SUM ÷ COUNT
```

There are a few problems with the above description: we need another memory cell in steps 3, 4, and 6; and "Repeat" isn't one of the operations we may use. The rule for reading a number off a data list specifies that the number be stored in a "named memory cell." (We refer to "the number," not a memory cell name.) Also, the algorithm doesn't explicitly stop, and we must consider what it would do under *general* circumstances: if the data list is empty, has only one value, has two values, etc. Modifying our version to insert these corrections, we get:

```
1. Set SUM and COUNT to 0
2. Repeat steps 3-5 while there is data left
3.    Read the next number off the data list, storing it in X
4.    Add X to SUM
5.    Add 1 to COUNT
   End of repeat
6. Compute SUM ÷ COUNT, storing the result in AVG
7. Write "The average of ", COUNT, " data values is ", AVG
8. Stop
```

ALGORITHM EXAMPLE 2–7

When, (as in step 7) an algorithm has quoted material in a write instruction, it means the words are to be written just as they are. Everything else (COUNT and AVG) must be a *variable* name. (Quotation marks indicate that words such as "The" and "average" are not variable names but characters to be written out verbatim.)

Now let's look at the four steps in general problem solving to see where they were used in solving the averaging problem.

Step 1 (Understanding the Problem). Using heuristics, we understood the problem and the required result. (Do you really understand the ramifications of sticking to the rules of the game?)

Step 2 (Devising a Plan). We devised an initial plan (sum them up and divide), then restated the plan with only the allowed operations.

Step 3 (Carrying Out the Plan). We haven't done this step, so let's try. Let the data list be the numbers 2, 5, 8. First, set SUM and COUNT to 0. Read the next number off the data list. In this case, that would be the first number, and so the value of X becomes 2. Add the value of X (2) to SUM (0) getting a new value for SUM: 0 + 2 = 2. Add 1 to COUNT (yielding a value of 1 + 0 = 1 in COUNT). Repeat steps 3-5 of the algorithm.

Read the next number, getting 5. Add it to SUM, yielding SUM = 2 + 5 = 7. Add 1 to COUNT, yielding COUNT = 1 + 1 = 2. Again, repeat steps 3-5.

Read the next number, getting 8, and add it to SUM, yielding SUM = 15. Add 1 to COUNT, yielding COUNT = 3.

Since there are no more data, go to step 6. Compute AVG = SUM/COUNT = 15/3 = 5. Write out the sentence: "The average of 3 data values is 5."

Finally, stop.

That seemed to work, and we got the right number as a result; but the third step in problem solving also asks us to consider special cases. What if the data list has no numbers? ("What a stupid question," you might say; but when you're dealing with a machine, anything can go wrong. Remember Murphy's Law!) If there are no data, the third step would go awry because there is no "next number" to be read from the data list. The computer will stop, or "abort," your program if you try to read something that's not there. (Even if the computer did not abort your program, it would complain when, in step 6, you direct it to divide by COUNT, which would be zero.) We have changed the algorithm so the computer will stop before it tries to read the first number, if there are no data.

Step 4 (Looking Back). This step in problem solving is retrospective, because every problem has many solutions (some more efficient or elegant than others). The averaging problem has a straightforward solution (and is the one we used); can you think of others?

2.7. THE NEED FOR GENERAL ALGORITHMS

At this point, we need to talk about general ideas behind algorithmic problem solving. The first idea is to describe a procedure or process that, when followed from an initial state or starting point, uses the data to compute the result. Since it takes effort to design or plan an algorithmic solution of a problem, we should write the algorithm in such a fashion that it will work for a variety of data lists.

For example, our algorithm for finding the average of a list of numbers will work for a list of any size (as long as it has at least one number). We could write it to work for only three numbers in the list, or only for the numbers 2, 5, and 8. But this, of course, would be a waste of time: how often do we need to find the average of three numbers or (even worse) of the numbers 2, 5, and 8?

The second idea in algorithmic problem solving, then, is to make the solution as general as practical—not so general that it solves the world's problems, but not so specific that it is useful in only a few instances.

The third idea is that the solution work for all cases it is designed to handle. For example, it is possible, to write an algorithm to find the average of a list of numbers that works only if the count is even and fails when the count is odd, or vice versa.

The fourth idea is to write pieces of algorithms that we can use in the solution of other problems. In this sense, averaging a list of numbers may be only a *module* in (or a *part of*) the solution of a larger problem. This will be a very important consideration as we write larger and larger programs. The more you can use solutions from previous programs, the less you must write from scratch. The more modular your programs, the easier to use modules from other programs.

2.8. SUMMARY

In this chapter we introduced a general four-stage plan for solving problems, and at each stage a number of questions were asked. These questions serve as heuristics to aid us in our mental search for a solution.

We discussed the necessity of naming memory cells so that we can refer to them. We've already named the actions by their usual names: if, repeat, add, store, recall, write, read, etc.

We partially solved the problem of averaging a list of numbers. We discussed the four stages of problem solving and tried to show how each applies to the example.

We again addressed the need for writing general algorithms, and gave four guidelines: they must be procedural; as general as practicable; complete; and modular.

2.9. EXERCISES

2.01. How could you modify the plan to count cars (in this chapter) so that it is an algorithm? Restate the plan as an algorithm. Be sure to remedy the two objections in the text.

2.02. Is the plan of action in Example 2-3 an algorithm? Briefly and concisely, justify your answer.

2.03. If step 8 in Example 2-3 were deleted, would the plan of action be an algorithm? Give your reasons.

2.04. Trace the plan of action in Example 2-3 using the value 4. What is the output? Try tracing it again, using the value 900 ($30 \times 30 = 900$). What is the output? Can you say anything about the accuracy of the plan? Can you suggest a way to improve it?

2.05. Change (modify) the square root algorithm so that the number of passes through the loop depends on the contents in s. In other words, repeat the loop until the result is "good enough." (What's "good enough"? How can we stop the algorithm if we don't repeat the loop a predetermined number of times?)

2.06. Rewrite the final version of the averaging algorithm to eliminate an objection raised in the text; that is, make it handle empty data lists.

2.07. Write an algorithm to find the smallest number in a list of numbers. (Hint: think about single elimination tournaments.)

2.08. Write an algorithm that counts the number of times the first number in a data list occurs.

Sample Input	*Sample Output*
5	The number 5 occurred 2 times.
1	
5	
3	
6	
3	
0	

Be sure the algorithm writes something appropriate (like "The data list was empty.") if there is nothing in the list.

2.09. Write an algorithm to count the number of values in the data list that are larger than the first value in the list. Do you notice any similarities between this problem and the previous problem? What are they?

Sample Input	*Sample Output*
3	There are 2 numbers greater than 3.
1	
5	
3	
6	
3	
0	

2.10. An algorithm that will write out the length of a list of numbers can easily be produced by modifying the averaging algorithm. Indicate these modifications and write the new algorithm.

3
Algorithmic Structures

Rule V
Method consists entirely in the order and disposition of the objects towards which our mental vision must be directed if we would find out any truth. We shall comply with it exactly if we reduce involved and obscure propositions step by step to those that are simpler, and then starting with the intuitive apprehension of all those that are absolutely simple, attempt to ascend to the knowledge of all others by precisely similar steps.

Rule VI
In order to separate out what is quite simple from what is complex, and to arrange these matters methodically, we ought, in the case of every series in which we have deduced certain facts the one from the other, to notice which fact is simple, and to mark the interval, greater, less, or equal, which separates all the others from this.

Descartes: *OEuvres*, vol. X, pp. 379 and 381; "Rules for the Direction of the Mind"

3.1. AN INSTRUCTION REPERTOIRE

Here we shall summarize, and be more explicit about, instructions in writing algorithms for a computer to execute. To do this, we must have a good idea of what a computer is. Although we present a fuller description in Chapter 5, we describe its essential points below.

A computer has three major units: its central processing unit (CPU), memory, and input/output system (I/O). Each of these units is connected to the others, so that our instructions affect all three.

The I/O system is the hardware that allows a program to move information between its memory and the outside world (i.e., print on a printer, read the characters typed at a terminal keyboard, read lines from a disk file, etc.). Here we employ other programs to prepare an **input list** for our algorithm to use as data. The input list, for our purposes, is a sequence of values (numbers, characters, etc.) that we give to an algorithm via a terminal, disk file, or other means of data input. Similarly, the algorithm produces an **output list** that the I/O system writes to a terminal, on a printer, or into a file. In an algorithm, we can give two instructions to the I/O system:

```
read name, name, name . . .
```

and

```
write name, name, name . . .
```

The command read x causes the I/O system to get the next item of data from the input list and put it in the memory cell we named x. Whatever was in x before the computer executed the read x command is gone and is replaced by the new value from the input list. The input list is advanced automatically by the I/O system, so that the next time we read a value we don't reread the same value, but the value *after* it. This means we cannot read a value a second time: *Once a value is read, it is gone from the input list.* Of course, we have a copy of the value in the memory cell we supplied with the read instruction; so there is no need to reread any value.

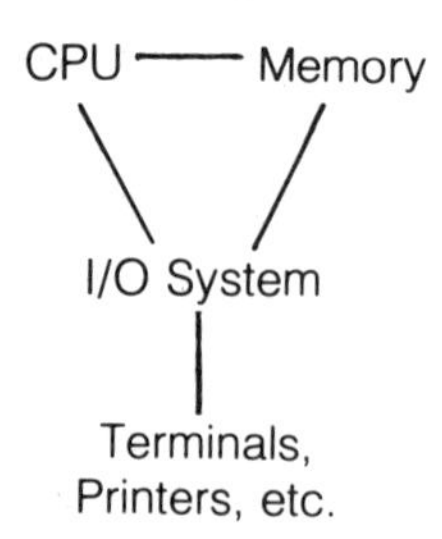

FIGURE 3–1 Major Computer Components

When we say read x, y to read in two values, the first value on the input list is put into x and the second value into y. We can read in any number of values in one read instruction by putting as many memory cell names in the read instruction as data values.

The last point we must consider with the read instruction is what happens if the computer executes a read instruction when the input list is empty (either to start with or because all the values have been read). In this case, most computers stop your program and tell you they can't proceed. It is an error, on our part, because our algorithm instructed the machine to do something impossible: read a nonexistent value. The correct procedure avoids such situations by asking the computer if there *is* any data before we instruct it to read.

You are wondering: "If the computer can tell when the input list is empty, why is it an error to read at that point?" The difference is between telling the computer to do something it *can* do and something it *cannot* do. When we ask if the input list is empty, we ask the computer to do something it *can* do: tell us if the input list is empty or not. Telling the computer to read a value, when there is none to read, is telling it to do something it *can't* do.

What *should* it do? If it doesn't cause an error, our algorithm will not work correctly, since the read should have obtained a value we need to continue our calculations. Obviously, that value is not available (*no* values are available); so our algorithm would be duped by the I/O system. We assumed we would receive a value, but we didn't. The algorithm, having no way to tell that something went wrong, will erroneously continue to execute, producing incorrect output (probably). Therefore, executing a read with no data available, is treated as an error.

The command write x causes the I/O system to write (or print) the contents of the memory cell we called x on the output list. For our purposes, we consider that each write instruction produces one line of output. If we assume x has the value 5 and y has the value 13.75,

```
write 'There are ', x, ' values totaling ', y
```

will produce one line of output:

```
There are 5 values totaling 13.75
```

The quoted characters are called **character string constants** and are written out exactly as given. The commands

```
write 'There are ', x
write 'values totaling ', y
```

will produce two lines of output (one for each write):

```
There are 5
values totaling 13.75
```

Programming languages handle the way numbers and characters are written on a page or terminal differently. In particular, there may be differences in the way numbers are written out: with extra blanks, in exponential form (e.g., 0.13750000E+02 instead of 13.75), etc.

The memory of a computer is a collection of memory cells, each of which can hold one integer number, one real number, one character, or one value of true or false. In Chapters 6 and 8 we will be more specific about the differences among integers, reals, characters, and true/false, and about aggregations of these and what they mean to a computer. Until then we can assume the first statement is true and can use the common understanding of integer, real, character, and true/false.

The instructions of a program are also stored in memory. Thus everything about a program is in the computer's memory.

Without memory, algorithms could not accomplish anything. The I/O system interacts with memory, via read and write instructions, and we use names to refer to particular memory cells. Memory cells can be examined or changed, just like the memory in most hand-held calculators. (Even if your calculator doesn't have a component called *memory*, it has at least one memory cell which holds the value in the display.) Examining the content of a memory cell retrieves the value; it does not clear or change it in any way. The write instruction copies the value in a memory cell to the output list.

To repeat, the value in the memory cell is not changed and can be examined again and again. *Changing* the contents of a memory cell places a new value in the cell and erases the old value. The read instruction changes the value in a memory cell by "substituting" the value of the next data item in the input list.

The CPU consists of the hardware to do arithmetic and logical operations. It also contains the hardware that decides which instruction to execute next. Except for read and write, all other instructions are carried out by the CPU. For our purposes, we need three more instructions: one to set a memory cell to the value of an arithmetic or logical expression; one to choose between two courses of action, based on the value(s) of one or more memory cells; and one to repeat a group of instructions.

Languages that are designed for use with computers, **programming languages**, usually have rigid rules that specify punctuation, the order of words, and so on. These rules define the **syntax**. However, we are more interested in the *meaning* of actions than in the *language syntax*. In algorithms, consequently, we can say what we want to do in any understandable way. We must be sure, nonetheless, that we limit ourselves to things the computer can do; otherwise our algorithms are useless.

The first instruction, to set a memory cell to the value of an arithmetic or logical expression, can be said different ways: For example,

Set x to 0
Put the value 0 into x
Make the value of x a 0
Store 0 into x
etc.

All mean the same thing. They are examples of an **assignment** instruction, in which the memory cell is *assigned* a value. In the examples above, the assignment changed the contents of x to the constant 0.

We can also use any arithmetic operators:

Set x to x+1
Increment x by 1
Add 1 to x
Take x, add 1 to it, and put the result back into x

Again, all of these statements mean the same thing. Set x to x+1 means Set the contents of the memory cell named x to the current contents of the memory cell named x plus 1. In all that follows, we assume the name of a memory cell stands for its content.

An expression is either a constant value (such as 1), or a variable (such as x), or a value obtained by computation (such as x+1). For the moment, we allow only the **arithmetic operators**:

+ addition
− subtraction
* multiplication
/ division (quotient with fractional remainder, e.g., 1/3 is 0.33333333)
÷ integer division (quotient without remainder, e.g., 4/3 is 1)

For consistency, we will use the form

set name to expression

most of the time. When another form seems more natural, we will use it instead. The idea is to say what we mean as clearly as possible—even though the concept of assignment is always the same, regardless of the different forms of expression. There is one way to do assignment in a computer programming language, and its syntax is strict. (As we said at the end of Chapter 1, an algorithm must be translated into a programming language so that we can run it on a computer.)

The second instruction—to choose between two courses of action, based on the value(s) of one or more memory cells—is referred to as **if-then-else**. It changes the natural sequential order (one after another) of executing instructions. The if-then-else allows us to select one of two groups of instructions for execution. Generally, it is written as

```
if some condition
  then   {if the condition is true}
    do this group
    of instructions
  else   {if the condition is false}
    do this group
    of instructions
end if
```

ALGORITHM EXAMPLE 3–1

The choice between the then and the else parts is based on "some condition," which is a question about the values of various memory cells or data on the data list. The questions we can ask about memory cells are about equality (or inequality), greater than (or equal), and less than (or equal).

For example, if NUM contains an integer and we need the absolute value of NUM, we could get it by saying

```
if NUM < 0
  then
    set NUM to -NUM
  else
    set NUM to NUM
end if
```

ALGORITHM EXAMPLE 3–2

But the else part is unnecessary; it doesn't do anything—and there are many cases like this. We need to do something if a condition is true, but nothing if the condition is false. It is reasonable to drop the else part under these circumstances:

```
if NUM < 0
  then
    set NUM to -NUM
end if
```

ALGORITHM EXAMPLE 3–3

The end if serves (in both cases) to show where the statements in the if instruction end. In the if-then-else version, the end if shows where the else part stops. In the if-then version, the end if shows where the then part stops, and also that there is no else part. (In the if-then-else, of course, the else shows where the end of the then part is as well as introducing the else part.)

After the if has selected the then part or the else part, only one, not both (based on the value of the condition), of that group of instructions is done and the natural sequential order resumes at the instruction that follows the end if.

The third and last instruction—to repeat a group of instructions—is called the **while loop**. This is one of two instructions we use to repeat groups of instructions; but for the moment, the while will suffice. The while loop, like the if-then-else, changes the natural sequential order of instruction execution. This loop repeats the instructions it controls *while* a condition is true. Repetition stops when the condition is checked and found to be false. Generally, the while loop is written

```
while some condition do
  group of
  instructions
end while
```

ALGORITHM EXAMPLE 3–4

The meaning of this while loop (what we tell the machine to do) is

1. Check the condition (as in the if instruction)
2. If the condition is false, stop the loop and continue sequentially after the end while
3. If the condition is true, execute the group of instructions; and then repeat this entire procedure, starting with step 1

The conditions in a while loop are the same as those that can be tested in an if-then-else—but, it is possible that the group of instructions in the while loop will not be executed. If the condition is false when it is tested the first time, the while loop is finished and the group of instructions is ignored.

It is important to realize that the if-then-else and the while loop are single instructions, even though they contain (or enclose) groups of other instructions. They are single instructions not only because they aggregate groups into a unit but also because they implement a single concept in algorithm design. As we proceed, we will encounter groups of instructions that implement important algorithmic concepts: standard ways to read in data to find the sum, average, minimum or maximum of a list of data values, etc.

We call these groups of instructions *modules* because most algorithms are built upon basic concepts implemented not by single instructions but by instruction *modules*.

3.2. EXAMPLE: FINDING THE BIGGEST

This section shows a use of the algorithmic structures described in the preceding section. Finding the biggest or largest in a list of numbers may seem trivial, but if we "play by the rules" we will find it's more complicated than we thought. One of the best reasons for studying algorithmic problem solving is to thoroughly understand how we, as humans, solve problems—to understand our assumptions and shortcuts as we hone our problem-solving abilities.

Given a list of numbers, find the largest number. If we are given this list,

```
    5
   −3
10000
    7
   10
    8
   −4
```

we immediately see the answer, 10000, because it "sticks out." This solution is dependent on the way we write numbers and the fact that a four-digit number must be larger than a three-digit number. We are trying to be general, and the "sticks out" technique doesn't always work. Try it on this list:

```
9873
9871
9876
9870
9872
```

As you see, a number does not "stick out," and the numbers are so similar in their digits that care must be taken to find the largest. Can you suggest another technique or procedure to find the largest number?

"Well, you sort of scan down the list and look for a big one and then see if it's bigger than all the others." This "procedure," though common, is faulty in several respects. One reason is that we cannot "sort of" do anything in an algorithm. We do it or we don't. Another reason is that if we "look for a big one," we might choose incorrectly, especially if the list is long and we can't examine it in a glance.

Moreover, we don't say how to decide whether a particular value is a "big one" or how to pick "the big one" out of many big ones. If we choose a number, then have to go through the whole list to see if it *is* the biggest, we're inefficient. Instead, we must find the biggest value by looking at each number only once. Finally, we commit a subtle error by checking to see if the "big one" is "bigger than all the others." The problem statement does not imply that the biggest number will be *unique*—that is, that there won't be several just as big. Of course, they would all be equal (several occurrences of the same biggest value), but where there are two or more of the biggest value, the "bigger than all the others" check would fail. (The biggest value is not bigger than itself.)

The discussion above was to point out the importance of careful planning, using heuristics, and adhering to the rules (legal instructions only) in writing algorithms. (We are not being picky.) The computer, unfortunately, is only a machine and can do only certain things: *what* it is told to do in the *way* it is told to do it. If we want its "cooperation," we must abide by its terms.

Again, writing a plan for finding the largest number in a list, let's follow the problem solving stages described in Chapter 2. First, do we understand the problem? The data? The result? (The answers, we hope, are yes.) Second, can we devise a plan? Do we know a related problem or similar problem for which we already have a plan? If you are familiar with competition in sports or other games, the answer is yes. A single-elimination tournament is the same sort of problem: Given a champion and a field of challengers, who becomes the new champion?

In this sort of tournament, the champion meets a challenger; the result is that the champion retains the title or the challenger wins, to become the new champion. The winner of any match during the tournament is presumed able to surpass all previous challengers (or champions). To remain champion, the current champion must face, and defeat, all challengers. At the end of the tournament, when there are no more challengers, the current champion is declared the winner.

To translate this into an algorithm that deals with a list of numbers, we recognize the first value in the data list as the "seeded" champion and the remaining values as the field of challengers. Given the data list

```
 5
-3
 5
 7
10
 8
-4
```

we get the results:

Seeded Champion: 5

Current Champion	*Challenger*	*New Champion*
5	−3	5
5	5	5
5	7	7
7	10	10
10	8	10
10	−4	10

Overall Winner: 10

If we write this as an algorithm, we need two memory cells. One cell holds the "champion" value and the other the value that corresponds to "challenger." The algorithm is

```
Start Algorithm
  read, champion   {get seeded champion value}
  while there is data left do   {repeat these instructions}
    read, challenger
    if challenger > champion
      then
        set champion to challenger   {a new champion}
    end if
  end while
  write, "The largest value is: ", champion
End algorithm
```

ALGORITHM EXAMPLE 3–5

In the third stage of problem solving we implement the plan, to see if it works. (This is sometimes called **tracing** or **playing computer**.) Starting at the beginning, the memory cells, or variables (champion and challenger), have unknown values. The data list is 5, −3, 5, 7, 10, 8, −4. (See Figure 3–2.)

As we see, the algorithm produced the correct result. Make another data list and trace the algorithm.

In the champion/challenger trace, we observe several things. We may change the value of a variable either by reading a data value into it or by *copying* (assigning) a value from another variable. Unless we change a variable by reading into it or assigning it a value, it does not change (even though we inspect it.) When a data value is read, it is *consumed* (i.e., once it is read into a variable, it cannot be read again), and the data list is shorter by one value. Conditions in the while loop and the if-then are checked only when it's their turn; they are not constantly monitored.

	Values after instruction is executed		
Instruction	Champion	Challenger	Data list
Start	?	?	5, −3, 5, 7, 10, 8, −4
read, champion	5	?	−3, 5, 7, 10, 8, −4
while there is data	condition is true		
read, challenger	5	−3	5, 7, 10, 8, −4
if challenger > champion	condition is false		
end if			
end while			
while there is data	condition is true		
read challenger	5	5	7, 10, 8, −4
if challenger > champion	condition if false		
end if			
end while			
while there is data	condition is true		
read challenger	5	7	10, 8, −4
if challenger > champion	condition is true		
then set champ to chall	7	7	10, 8, −4
end if			
end while			
while there is data	condition is true		
read, challenger	7	10	8, −4
if challenger > champion	condition is true		
then set champ to chall	10	10	8, −4
end if			
end while			
while there is data	condition is true		
read, challenger	10	8	−4
if challenger > champion	condition is false		
end if			
end while			
while there is data	condition is true		
read, challenger	10	−4	empty
if challenger > champion	condition is false		
end if			
end while			
while there is data	condition is false		

write, "The largest value is:", champion

output: The largest value is: 10

stop

FIGURE 3–2 Tracing an Algorithm

The key element in this algorithm is that the while loop reads the challengers in, one at a time, and matches the current challenger against the current champion. The repetition of the read and the if statements allows the lengths of the data lists to vary from one list to another. The if statement compares the challenger with the champion and determines whether the challenger wins. If the challenger wins, the value of the challenger variable is copied (assigned) to the champion variable, at which point the values of the two variables are equal. The old champion (as in real life) is forgotten and never heard from again.

The fourth stage of problem solving—looking back—asks us to reflect on our solution. Could we have done it a different, perhaps better, way? Can we use this algorithm as a module in solving another problem?

The answers are yes and yes. We can certainly use the algorithm in the solution of another problem; many problems entail finding the largest or best, etc. We can "customize" the algorithm to find the "largest" in solving problems that are more complex, such as the best per gallon mileage when you know the gas used and the miles traveled by various cars. To customize an algorithm, we rename the variables in keeping with the names in the rest of the algorithm, so they "make sense" in that context. Also, we might have to add instructions to compute other results. For example, to find the MPG for a given car, we have to divide miles traveled by gallons of gas used before we could use the result as the challenger value in the find-the-largest algorithm.

Also, we can use the find-the-largest algorithm in a different way. Finding the largest and finding the smallest are very similar tasks—so similar, in fact, that to change the find-the-largest algorithm into a find-the-smallest algorithm, all we do is change

```
if challenger > champion
```

on line 5 to

```
if challenger < champion
```

and change the write statement to print "The smallest value . . ." instead of "The largest value . . ." Trace the modified algorithm (3-6) to assure yourself that these simple changes work.

```
Start Algorithm
  read, champion   {get seeded champion value}
  while there is data left do   {repeat these instructions}
    read, challenger
    if challenger < champion
      then
        set champion to challenger   {a new champion}
    end if
    end while
```

```
    write, "The smallest value is: ", champion
End algorithm
```

ALGORITHM EXAMPLE 3–6

There are other ways as well to find the largest or smallest, such as another slight (but frequently used) variation of the champion/challenger technique. First, however, consider this customization of the find-the-largest algorithm to find the car with the best MPG:

```
Start algorithm
  read, gallons, miles
  set BestMPG to miles/gallons
  while there is data left do
    read, gallons, miles
    set NewMPG to miles/gallons
    if NewMPG > BestMPG
      then
        set BestMPG to NewMPG
    end if
  end while
  write, "The best MPG is: ", BestMPG
End algorithm
```

ALGORITHM EXAMPLE 3–7

We had to make five changes to customize the algorithm. Two of these changes simply renamed the variables, Champion to BestMPG and challenger to NewMPG to make more sense in the new context. Another simple change set the write statement to print a more meaningful message: "The best MPG is:" instead of "The largest value is:". These two changes involve the two read statements in the original algorithm.

Now we read two data values, gallons and miles, and compute the mileage before we continue. The mileage value becomes the challenger value, since we are looking for the best miles per gallon. We have to make this change in both places where the read appears: that is, we have to duplicate the computation of miles per gallon.

This duplication is bad for two reasons. If we ever need to change the way we compute miles per gallon, we must change it in both places. It is best not to trust to memory for such things, if we can avoid it—that is, if we can compute MPG in only one place instead of two. Also, if the computation is more involved than that for MPG, we may have to duplicate many instructions. This is inefficient, if we can avoid duplication by doing the computation in only one place.

It seems that we should redesign the algorithm so that we read data and do associated computation at only one point. If we remove the read and computation from one of the two places where they currently stand, we have to remove the first occurrence. If we remove the occurrence inside the while loop (the second occurrence), we won't read any more data and the loop will go on forever (no data is read, so the data list doesn't shrink; so data is always left). Therefore we must choose the occurrence outside the while loop.

If we eliminate the first read, however, we remove the statement that gives champion and BestMPG their initial values. This leads to a problem later, when we check challenger against champion, in

```
if challenger > champion
```

Champion has no value; so we can't know the result of the comparison, and the algorithm will no longer produce predictable results.

We can sidestep this issue by initializing champion (or BestMPG) to an absurdly small value. For example, we could initialize BestMPG to zero:

```
Start algorithm
  set BestMPG to 0.0
  while there is data left do
    read, gallons, miles
    set NewMPG to miles/gallons
    if NewMPG > BestMPG
      then
        set BestMPG to NewMPG
    end if
  end while
  write, "The best MPG is: ", BestMPG
End algorithm
```

ALGORITHM EXAMPLE 3–8

This design works because any authentic MPG is positive, and any positive MPG is larger than zero MPG. So the zero value is displaced by the first authentic (positive) value to come along.

This design has one disadvantage, but two advantages over the first design of the find-the-largest algorithm. It works only when we can start champion at an "absurd" value. When the second design is inappropriate or awkward, the first design should be used. The advantages of the second design are that we've simplified the algorithm by eliminating one read and one computation, that *this algorithm will also work for an empty data list.*

If we give an empty data list to the first design, the computer rejects it when we tell it to read, Champion. There is nothing to read. If we give an empty data list to the second design, the while loop stops (so to speak) before it starts because there is no data the first time through; and so it prints "The best MPG is: 0.0." (What it prints is incorrect, but at least we weren't rejected by a machine.) It prints an MPG of 0.0 because the value of BestMPG was initialized to 0.0 and never changed. It is easy to fix the second design so that it behaves rationally when it's given an empty data list:

```
Start algorithm
  set BestMPG to 0.0
  while there is data left do
    read, gallons, miles
    set NewMPG to miles/gallons
    if NewMPG > BestMPG
      then
        set BestMPG to NewMPG
    end if
  end while
  if BestMPG > 0.0
    then
      write, "The best MPG is: ", BestMPG
    else
      write, "There were no data to be read."
  end if
End algorithm
```

ALGORITHM EXAMPLE 3–9

We check to see if BestMPG is still 0.0 when we get out of the while loop. If it is not, some value of NewMPG must have been greater than the original (zero) value of BestMPG. This implies at least one pair of data, and we write out what must be the correct result. If BestMPG is still 0.0, there must have been no data (at least no *valid* data), and we write a message to that effect.

3.3. CHECKING FOR VALID DATA

Since the computer is completely subservient to instructions, we take the responsibility for ensuring that our algorithms anticipate every eventuality. We must be sure we do not instruct the machine to attempt something impossible (read nonexistent data, divide by zero, etc.). We must also be sure that values over which we have no control are checked for validity before we use them in computations.

Data values are such values; they may be mistyped, misread, etc., or a key may have stuck on the terminal, or the wrong data file may be used. Any number of things might cause bad data to be presented to our algorithm.

Let us assume we have translated our best-MPG algorithm into a program and made it available to everyone who uses the computer. Let's also assume that someone uses it and gives it the following data:

```
-5  -200
10   390
 8   312
12   470
```

There are two obviously bad data values: it is impossible to use −5 gallons of gas and to travel −200 miles. Negative numbers make no sense in this case. As it happens, our algorithm will report a best MPG of 40, which corresponds to −5 gallons and −200 miles. Our algorithm is wrong, and we should have reported the −5 and −200 as invalid data. When we discover invalid data, we must decide what to do (other than report it, which we must always do).

We could have the algorithm do one of several things if it encounters invalid data: print a message and stop. Or we could have the algorithm print a message, ignore the invalid data, and continue with the remaining data. Or we could have the algorithm print a message, substitute some valid value for the invalid one and continue. And so on. Under *some* circumstances, each of the three strategies works.

The first strategy, to print a message and stop, is used when valid data is absolutely required for the algorithm to continue. If any value is invalid, however, the algorithm must quit. For example, assume that the first number in a list indicates how many more values we are to read. If this first value is negative, we have no idea how to proceed, and must stop the algorithm.

The second strategy, to print a message and ignore invalid data, is used when the data is independent of all the other data. For example, if the problem were to print out MPG, given gallons and miles, without looking for the *best* MPG, an error in one pair of gallons and miles would make no difference in the MPG reported for the next pair, and so could be ignored.

The third strategy, to print and substitute, is used when there is an accepted "standard" value to use as the substitute value. For example, suppose we wrote an algorithm to help a parking-lot attendant figure how much to charge. The attendant would enter the time a customer entered the lot and the current time. If either or both times were in error, we could substitute the maximum charge for the charge obtained in the usual way.

In our dilemma over the negative gallons and negative mileage, which solution should we use? To print an error message and stop seems too harsh, but to print an error message and ignore the invalid data might lead to an incorrect answer.

In such a case, we consider the user of our algorithm. If we print a message and stop, the user must correct the data and rerun the whole thing—during which the algorithm may find more invalid data and stop again (or finish).

If we print and ignore, the user is warned of invalid data and that the answer might be wrong. In this case, it is easy to check by hand to see if the bad data would affect the result if it were corrected and replaced. If there are multiple invalid data, all are identified in one execution of the algorithm, making it easier to correct and rerun (if desired).

To amend the best-MPG algorithm to detect invalid data, we check gallons and miles with an if instruction, because we must choose between two courses of action: one for valid and the other for invalid data.

```
Start algorithm
  write, "The data is:"
  write, "Miles Gallons MPG"
  set BestMPG to 0.0
  while there is data left do
    read, gallons, miles
    if (gallons > 0.0) and (miles > 0.0)
      then
        set NewMPG to miles/gallons
        write, miles, gallons, NewMPG
        if NewMPG > BestMPG
          then
            set BestMPG to NewMPG
        end if
      else
        write, miles, gallons, "— Data must be positive."
        write, " These data ignored."
    end if
  end while
  if BestMPG > 0.0
    then
      write, "The best MPG is: ", BestMPG
    else
      write, "There was no data to be read."
  end if
End algorithm
```

ALGORITHM EXAMPLE 3–10

In addition to printing the data and the message when the data was in error, we print the valid data. The two write instructions at the start of the algorithm print headings, under which we write the data as we read it in. This is a good practice to learn. When we print the error message, we should include: (1) the data we think is in error, (2) why we think it's in error, and (3) what we've instructed the algorithm to do about it. In this example, we write both miles and gallons, one or both of which is in error. We also write "Data must be positive." to indicate that negative or zero values are unacceptable. Finally, we write "These data were ignored." to indicate that these values were not used.

If we trace this last version of the best-MPG algorithm, using the data

```
-5    -200
10     390
 8     312
12     470
```

the algorithm will generate this output:

```
The data is:
Miles        Gallons      MPG
-200           -5         —Data must be positive.
                            These data ignored.
 390           10         39.00
 312            8         39.00
 470           12         39.17

The best MPG is: 39.17
```

Trace the algorithm; understand why it works; and remember these techniques. We will need them to design other algorithms.

3.4. A NON-NUMERICAL EXAMPLE: THE COMBINATION LOCK

The earlier problem in Chapter 3 (averaging a list of numbers) was primarily numerical. Since computers are built into many machines to control their actions, let's consider a simple problem in the area known as **process control**.

Assume you've been hired by ACME Lock and Safe Company, and your first job is to build a machine that will check a lock to see if its combination (printed on the tag) really opens it. In other words, you're to build a machine to do quality control for the company.

Also assume that the company has assigned a group of engineers to build a machine that can manipulate a lock (i.e., turn and read its dial). The control mechanism, however, is a computer—the part of the machine that determines what it does next. Unfortunately, the engineers don't have time to program the computer, and that's where you fit in: you write an algorithm. The algorithm to "solve" a combination lock is printed on the card on the lock's shank—but you're dealing with many locks, and the solution (combination) on one card won't (or shouldn't) work for another lock. The primary consideration, though, is that all locks have the same solution *in general*. You can *parameterize* the solution (algorithm) by calling each of the three numbers in a combination by names, say i, j, and k, and express the algorithm like this:

Algorithm to Open a Combination Lock (with 60 numbers on the dial, whose combination is represented by i, j, and k.)

1. Turn right 2 full turns
2. Turn right to i
3. Turn left 1 full turn
4. Turn left to j
5. Turn right to k

ALGORITHM EXAMPLE 3–11

Step 1 brings the lock to an "initial state," so that the rest of the instructions will work properly. The other steps cause the lock to go from the initial state (ready to accept the first number) to three other states in succession (ready to accept the second number, ready to accept the third number, then open).

This doesn't help much if you think about a machine's doing it, since you don't know what the machine is capable of doing. So you must ask the engineers who built the machine what operations (instructions) it can perform, and they respond with this description.

The machine has a clamp that holds the lock and a rubber cup that fits over the knob and is attached to a motor. The machine can manipulate the lock in the following ways:

1. Turn the knob 1 click to the right
2. Turn the knob 1 click to the left
3. Compare the number at the top of the dial with one presented in an "instruction"

Furthermore, the machine can be given an algorithm (whose form we need to describe) and a three number combination. These are the rules of the game.

Now describe the form of the algorithm. If you were to limit your instructions to the basic set we've described for the machine, you would have a very long algorithm. The first step alone would expand into 120 instructions (we would have to turn the dial 120 clicks to the right to get two full revolutions since the dial has 60 numbers). Obviously, you need a loop that allows you to specify repetition of a group of instructions a predetermined number of times. Thus the first step in our unlocking algorithm (3-11),

```
turn right 2 full turns
```

can be rephrased in terms that are closer to the machine's basic operations:

```
turn the dial 120 clicks to the right
```

But this still is not what the machine understands. It wants to be told to turn the dial one click at a time. So let's say this:

```
set x to 1
while x ≤ 120 do
   turn the dial 1 click to the right
   increment x by 1
end while
```

ALGORITHM EXAMPLE 3–12

The purpose of this loop is to repeat the instruction "turn the dial 1 click to the right" 120 times. The variable x is used to count the number of times the loop is executed.

Most programming languages provide another mechanism to do exactly this sort of thing: repeat a group of instructions a predetermined number of times. Accordingly, we obtain the same effect as the while loop by using a **for loop**:

```
for x ← 1 to 120 do
   turn right 1 click
end for
```

ALGORITHM EXAMPLE 3–13

We write this structure for, followed by a variable name (called an **index variable** in this case), followed by a starting value; then by the word to, an ending value, the word do and finally an instruction. This can be written (for our purposes)

```
for INDEX ← START to STOP do
   instructions
end for
```

The end for, (like the end while) indicates the end of the instructions inside the loop (i.e., exactly how much is to be repeated).

The way the for loop is carried out or executed is rather simple. INDEX is a variable (x in this example) in which a value is stored, and this value begins at the START value (1 in this example). Now the value in INDEX is compared to the value STOP. If the INDEX is less than or equal to STOP, the instructions are executed and the INDEX is incremented by 1. At this point, the INDEX is again checked against STOP and everything repeats, until the comparison finds that the value of INDEX is greater than STOP, at which time the loop stops. Execution of the for loop can be restated algorithmically:

```
1. Assign the value START to INDEX
2. while INDEX ≤ STOP do
3.     Do the instructions.
4.     Increment the value of INDEX by 1
5. end while
```

ALGORITHM EXAMPLE 3–14

This looping structure allows us to repeat a set of instructions a predetermined number of times. INDEX allows the loop to determine when to stop (it begins at the START value and advances to the STOP value; its value changes by 1 each iteration through the loop, and so keeps track of the loop's iterations.

Let's return to the problem of opening the combination lock. The next step we must translate (from Example 3-11) is 2. Turn right to i. Although we can only turn the lock one click at a time, we turn many clicks in several ways. If we use the only control structure we know, we might say:

```
for x ← 1 to i-dial do
   turn right 1 click
end for
```

If dial is the number at the top of the dial, i-dial is the number of steps it takes to get from where we're at (dial) to i (the first number in the combination). But wait a minute. It *looks* like the instruction would work—but it doesn't work *in all cases*. It works correctly only if i ≥ dial (if i is greater than or equal to dial). If (say) i = 5 and dial = 30, then 5 − 30 = −25 and the for loop will do nothing (since the starting value is greater than the stopping value). This sort of mistake is called a **bug**, and finding and fixing bugs is called **debugging**. (Captain Grace Murray Hopper, U.S. Navy, says the original "bug" was a moth, caught in a relay of the MARK I computer [1944].)

(You should learn from this mistake. Never trust anything you're not absolutely certain of, and be wary of all you are certain of. You never really know until you see something work. Even then you can't be sure it will work *in every case*. Be sure the instructions you write do what you intend them to do for all the cases you will *ever* encounter.)

Back to our problem with i-dial. How can we fix it? One way involves an interesting mathematical expression (x modulo y is the remainder after x is divided by y):

```
(i-dial + 60) modulo 60
```

This yields the right number of clicks but is not very clear. (Check it out, if you're a mathematical type.) So let's try to do this in a more straightforward manner. The trouble is, we might get a negative number, subtracting dial from i. So we use the if instruction:

```
if i > dial
   then
      for x ← 1 to i-dial do
         turn right 1 click
      end for
   else
      for x ← 1 to (60-dial) + i do
         turn right 1 click
      end for
end if
```

ALGORITHM EXAMPLE 3–15

This is a check to be sure that i-dial is positive before we allow the for i ← 1 to i-dial do to execute. Then, if i-dial is not positive (more exactly, when i is not greater than dial), we execute the for loop in the else part of the if. The expression 60-dial+i gives the number of clicks to turn: 60-dial moves it to 60, then +i moves it that many clicks more. For example, if dial = 30, and i = 5, we need to move 35 clicks: 30 clicks to get from dial = 30 to dial = 60, and 5 more to get dial = i = 5.

Expecially in devising a plan, we are wary of complicated solutions to problems that are essentially simple. Our problems is that we're trying to use the for loop for everything and have ignored the other possibility, the while loop. As we've pointed out, the for loop is handy when we know *beforehand* how many repetitions are required. When we *don't* know beforehand, it is usually advisable to use a while loop.

The second step in Example 3-11,

2. Turn right to i.

can be accomplished by the while loop's repeating turn right 1 click as long as the dial ≠ i:

```
1. for x ← 1 to 120 do
      turn right 1 click
   end for
2. while dial ≠ i do
      turn right 1 click
   end while
```

The while loop checks its condition before it attempts to execute the loop body. If the dial is positioned at i at the end of step 1, step 2 merely checks the dial. Otherwise, it repeatedly turns the dial 1 click to the right, until it arrives at i.

Let's analyze what we did and why we did it. First, we asked two questions: "What state is the lock in?" and "What state do we want to put the lock into?" (These questions are echoes from our discussion of problem solving: "What are the initial conditions?" and "What do we need as a result?") In step 1 in the algorithm, we "cleared" the lock mechanism by turning it right two complete revolutions; thus the lock will "accept" the first number in the combination. The object of step 2 in our algorithm is to advance beyond this "cleared" state, to a state in which the lock can accept the second number of the combination. In short, we've moved the dial to i, the first number of the combination, and the remaining steps in the solution are variations of our previously discussed steps.

Step 3, 3. Turn left 1 full turn. is similar to step 1: to turn right two full turns. The instructions don't differ in a general sense (both turn the dial a predetermined number of steps) and thus are analogous. (Again, this is an echo from our problem solving discussion: Look for previously solved problems that are similar, and Look for analogous problems.) As we see, step 3 means that we start with the lock in the state it's in just after the input of the first number in the combination, then set it into a state so that it's ready to accept the second number. Operationally, this involves turning the dial one full turn left. Using a for loop (as in step 1), we get

```
3. for x ← 1 to 60 do
     turn left 1 click
   end for
```

The next step, 4. Turn left to j is analogous to step 2 (only the direction and stopping point are different). We state this using a while loop (as in step 2):

```
4. while dial ≠ j do
     turn left 1 click
   end while
```

(Students should analyse this problem for initial and final states to understand and thereby be able to describe the meaning behind the operational definition).

Finally, 5. Turn right to k is just like steps 2 and 4; so we can also state this with a while loop:

```
5. while dial ≠ k do
     turn right 1 click
   end while
```

The algorithm is finished and the machine should stop (also, the lock should be open). The algorithm to open (solve) the combination lock (given the combination i, j, and k) is

```
1. for x ← 1 to 120 do
      turn right 1 click
   end for
2. while i ≠ dial do
      turn right 1 click
   end while
3. for x ← 1 to 60 do
      turn left 1 click
   end for
4. while j ≠ dial do
      turn left 1 click
   end while
5. while k ≠ dial do
      turn right 1 click
   end while
6. stop
```

ALGORITHM EXAMPLE 3–16

3.5. EXAMPLE: EXTRACTING DIGITS FROM A NUMBER

Given a list of positive numbers, for each number in the list

```
read a number
write the number just read
while there are digits to print do
  remove the rightmost digit and put it in DIGIT
  write DIGIT
end while
stop
```

ALGORITHM EXAMPLE 3–17

First, let's solve the problem for a single number, say N, and to learn how to do this *in general*, we'll identify *patterns*. Given the number 195, how do we obtain the last (rightmost) digit? (It isn't fair to say, "Well, just look at it and that's it.") From second or third grade (or whenever you were first introduced to division), you might recall that division of integers results in a whole number as a quotient and another whole number as a remainder. If we divide 195 by 10 in this manner, we get a quotient of 19 and a remainder of 5; thus the remainder appears to be the last digit—but what about the next digit, the 9?

The quotient of the previous operation is 19, so that 9 is now the last digit. Moreover, 19 divided by 10 yields a quotient of 1 and a remainder of 9—so what about the 1? Again, the last quotient's last digit is the number we need; so 1 divided by 10 yields a quotient of 0 and a remainder of 1. Looking at the number, we know we are finished; but if we didn't (or couldn't) look, how might we tell whether we were done? It looks as though the quotient holds the key: As long as the quotient is greater than zero, there is more to go; when the quotient is zero, we are done.

Let's formulate an algorithm for this problem. The write and while instructions are already in our standard algorithmic form; however, "remove the rightmost digit and put it in DIGIT" is not. We must say how we will accomplish this, using only our usual instructions. This latter instruction can be "translated":

```
set quotient to number ÷ 10
set remainder to number − quotient*10
```

Remainder contains the digit we need. (There may be no easy way to ask for the remainder of a division; we assume you have to figure it out yourself.) The next step is straightforward: Print the remainder. The end while shows the end of the loop but how should the while work? Consider two things:

Are things ready for the next time through the loop?
When should the loop stop?

The answer to the first question is no: if we go back right away, without changing the number, we will get the same result as before.

Doing this example, we saw that, to get the second digit, we must work with the quotient we got the first time, and to get the third digit, we must work with the quotient we got the second time, etc. It appears that before we go back, we must replace the content of the variable that stands for the number read by the quotient of the previous operation: set number to quotient

The loop should stop—in answer to the second question—when the quotient becomes zero (as we saw when we did the example by hand).

At this point, we have refined the steps we started with:

```
Start algorithm
   read, number
   write, number
   while number > 0 do
      set quotient to number ÷ 10
      set remainder to number − (quotient * 10)
      write, remainder
      set number to quotient
   end while
End algorithm
```

ALGORITHM EXAMPLE 3–18

Each of these steps, moreover, is easily translated into a programming language.

Now we turn to the problem of handling a list of numbers, and hence to a sequence of computations like the one we just did. Handling a list of numbers, we use the standard loop structure:

```
while there is data do
   read a number
   use the number in computations
end while
```

ALGORITHM EXAMPLE 3–19

This yields the following algorithmic solution of the problem:

```
 0. Start algorithm
 1.    while there is data left do
 2.       read, number
 3.       write, number
 4.       while number > 0 do
 5.          set quotient to number ÷ 10
 6.          set remainder to number − (quotient * 10)
 7.          write, remainder
 8.          set number to quotient
 9.       end while
10.    end while
11. End algorithm
```

ALGORITHM EXAMPLE 3–20

Now let's try to convince ourselves that our algorithm works. Looking at the algorithm, we see we have names: number, quotient, and remainder. Given the input list, 159, 2, what will the program produce? We can trace the program by keeping track of the contents of all the names. (The numbers to the left of each instruction above are used below as step numbers.)

step	*value of:*	*number*	*quotient*	*remainder*	*output*
1	while ok				
2		159	?	?	
3					159
4	while ok				
5			15		
6				9	
7					9
8		15			
4	while ok				
5			1		
6				5	
7					5
8		1			
4	while ok				
5			0		
6				1	
7					1
8		0			
4	while quits				
1	while ok				
2		2			
3					2
4	while ok				
5			0		
6				2	
7					2
8		0			
4	while quits				
1	while quits				
STOP					

The output is 159 ← the original number
9
5
1
2 ← the original number
2

All appears to be okay.

3.6. SEEING THE FOREST IN SPITE OF THE TREES

> Our life is frittered away by detail. An honest man had hardly need to count more than his ten fingers, or in extreme cases may add his ten toes, and lump the rest. Simplicity, simplicity, simplicity. I say, let your affairs be as two or three, and not a hundred or a thousand; instead of a million count half a dozen, and keep your accounts on your thumb-nail.
>
> Henry David Thoreau, *Thoreau On Man & Nature*

Abstracting general concepts from details and using symbols in place of objects are very important in thinking algorithmically. Of the two principal types of abstraction in algorithms, *symbolic abstraction* is the more familiar. The type by which we understand the meanings of algorithms we call *structural abstraction*.

Symbolic abstraction deals with using symbols with which we can build symbolic models. The simplest kind of symbolic abstraction is **encoding**, that is, using a code to represent an object. Part numbers, stock codes, student numbers, Social Security numbers, etc. are encodings. A symbol is used to *represent* a thing.

For example, computerized student enrollment systems implement an abstract or symbolic model of the enrollment process. In real life, a student attends class in a certain room with a certain instructor. In the abstract model, a *student number* is associated with a *class number* and *staff number*. In addition, the class number is associated with a *room number*, a *class title*, etc.

Having used these models, we know it is difficult to maintain accuracy—that is, ensure that they accurately reflect the real world. (Many students have discovered this in painful ways.) We use symbolic abstraction when we formulate models of problems we want to solve.

Structural abstraction deals with the algorithmic solution of a problem. When we solve problems—even though we know beforehand that we will eventually need an algorithm—we usually do not state the solutions in the algorithmic structures we have used thus far. We think of the solutions in more general terms, more general such as "Find the average," or "Sort the names," etc. Each of these general steps can be stated in more and more detail, until we state them in terms of the sequential operations read, write, and assignment—together with the control structures: if-then-else and while. This is called **step-wise refinement**, because at each more detailed step we refine the generality of each instruction into more specific instructions.

The problem in this refinement process is losing sight of the forest (the general statement of the solution and the objective of our efforts) because we get too involved with the trees (the instructions by which we implement the general statements). Studies of programming projects indicate that we cannot keep all the details in even simple algorithms in our heads at one time. To write algorithms for large problems, it is necessary to state solutions in terms of their major facets or subproblems.

After the overall design is finished, (in the most general terms), we refine each instruction in more and more specific terms, including more detail at each step. This is sometimes called **divide and conquer**. We may not be able to write an entire algorithm immediately, but we can divide a large problem into subproblems, each of which we can conquer.

Structural abstraction, then, deals with dividing a problem into subproblems. As we study algorithms, we recognize common subproblems that can be restated directly in algorithmic terms. Because we can't always deal with all the details in many problems, it is important to learn how common subproblems are solved and how to recombine them to solve larger problems. Chapter 4 begins this introduction.

3.7. SUMMARY

At the beginning of this chapter we introduced an instruction repertoire, including

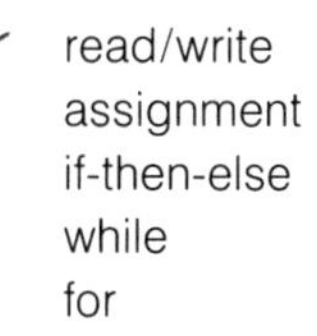

read/write
assignment
if-then-else
while
for

These are the five basic algorithmic structures (instructions) we must use when we write algorithms. We discussed a way of stating each instruction, and saw that the concept—not a particular statement—is important.

We wrote and rewrote an algorithm to find the largest number in a list of data. At each step, we consulted our list of heuristics and tried to find ways to improve the algorithm. We made certain, for example, that it would handle empty data lists.

We discussed invalid data and how to write an algorithm that handles such situations rationally. Also, we saw the importance of conceiving an algorithm from the user's viewpoint (rather than from the author's).

The combination lock problem demonstrates problem states and state-to-state transitions. To solve a problem, we need to determine

1. The *initial state* of things—or decide that we don't know, but can get a known initial state by doing something (like turning a dial two full turns right)
2. The set of commands to describe *operationally (procedurally)* how to go from one *state* to another
3. The *final state* of things (or result)

We showed the difference between what we mean to accomplish and the way we describe its accomplishment (i.e. an algorithm is an *operational description* of the solution, *given the proper starting conditions*). We also introduced the concept of *control structures*, which allow us to condense an algorithm (by using loops) and do things we otherwise couldn't (if-then-else; while loops, for loops).

A number of questions led toward a solution, such as "What are the initial conditions?" and "What do we need as a result?"—as well as such hints as "Look for previously solved problems that are similar" and "Look for analogous problems." (Bear them in mind whenever you solve problems, with computers or otherwise, since they help you focus on useful lines of thought.) *Parameterization* of a solution states the solution of a more general problem. The terms *bug* and *debugging* were introduced, with the warning that your instructions work for *all* cases, not just some.

Finally, symbolic and structural abstractions are important in straight-line algorithmic problem solving because they allow us to specify and manipulate a model and to organize necessary detail in a coherent fashion.

3.8. EXERCISES

3.01. Write an algorithm to find the worst MPG, given a list of gallons and miles. Use Algorithm Example 3.10 to start.

3.02. What happens if you give the data list -5, -3, -9, -1 to Algorithm Example 3.10? Trace the algorithm, using these data, and show the output.

3.03. Generalize the combination-lock solution for models that have various numbers on the face of the dial (the solution in the text is tied to 60 numbers on the face of the dial). A solution that requires a four-part (instead of a three-part) combination is *not* desired.

3.04. Write an algorithm to process the weekly payroll of Acme Publishing Company. For each employee, your program will compute gross pay, deductions, and net pay. Each week, APC collects the following information for each employee:

Social Security number
Hourly pay rate
Number of exemptions
Hours worked

On the data file are lines of data, and each line has four numbers, representing the information for one employee.

Your algorithm should compute

a. Gross pay: regular pay for the first 40 hours and time-and-a-half beyond that

b. Deductions: let G represent gross pay, T taxable pay, and E the number of exemptions

   1. Federal income tax withholding is defined as $T = G - 14.00 * E - 11.00$
      Withholding $= T * (0.14 + 0.00023 * T)$
   2. Social Security tax is 16.70 or 7.7% of G, whichever is smaller

c. Net pay: gross pay less all deductions

For each employee, your program should output (with labels) his/her Social Security number, gross pay, type of each deduction, and net pay on separate lines. Your algorithm should work for an arbitrary number of employees. If taxable pay is negative for an employee or total deductions exceed gross pay, there should be no deductions for that employee.

Sample Input

```
123456789    5.00  1  40.0
138532110    7.00  3  60.0
```

Sample Output

```
Social Security number   123456789
Hours worked                  40.0
Rate of pay                   5.00/ hour
Number of exemptions             1
Gross pay                   200.00
Federal withholding          31.54
Social Security tax          15.40
Net pay                     153.06

Social Security number   138532110
Hours worked                  60.0
Rate of pay                   7.00/ hour
Number of exemptions             3
Gross pay                   490.00
Federal withholding         105.10
Social Security tax          16.70
Net pay                     320.94
```

3.05. ABC Company buys bulk rope and cuts it to various lengths, to be sold in retail stores. As a quality control measure, samples of the cut rope are chosen and the length of each (in meters) is determined. A length is within tolerance if it is within 1% of the intended length (i.e., between 99 and 101%). The first line of data gives the intended length of samples. The remaining lines contain the lengths of an unknown number of samples, one to a line. Write an algorithm to check rope lengths.

(Note: ABC has just hired a new person to enter data into the file, and this person at times makes mistakes. Assume a length in the file is correct unless it is negative, then it is a mistake.

Sample Input

```
  36.5
  36.5
  36.0
  36.7
  38.0
 -36.5
```

Sample Ouput

```
Ropes should be 36.5 meters long.
  36.5 is acceptable
  36.0 is not acceptable
  36.7 is acceptable
  38.0 is not acceptable
 -36.5 is an illegal entry
```

3.06. Write an algorithm that reads in 2 integer numbers (say M and N) and prints out the first M numbers of the extended-Fibonacci series, starting at N. (This series is generated as follows: The first number is N; the second number is always the same as the first; each number after the second is the sum of the previous two numbers.

Sample Input   *Sample Output*

```
   7  1         Term number       Fibonacci number
                     1                   1
                     2                   1
                     3                   2
                     4                   3
                     5                   5
                     6                   8
                     7                  13
```

3.07. Read from a data file which contains a list of positive integers, to determine if each integer is even or odd. Also determine the largest odd and the smallest even integers. (Make sure your algorithm will produce reasonable results if there are no odd [or even] numbers in the file.)

Sample Input

```
 8
 1
 6
 5
 3
-2
```

Sample Output

```
8 is even
1 is odd
6 is even
5 is odd
3 is odd
The largest odd number was 5
The smallest even number was 6
```

As the sample data makes clear, the smallest even and the largest odd numbers are unrelated. The smallest even number may be larger than the largest odd number.

3.08. Write a algorithm to handle data as described below. Each line of input has a code in column 1 to indicate the type of computation required on the 4 integer values following it. Do not process a data line if any of the values lie outside the range 10 to 20. The codes are:

S - sum of the squares
C - sum of the cubes
A - average of the squares

Print the value of each computation together with an appropriate message (do not forget to echo print). If any of the values are invalid, print the data items and an error message.

Sample Input

```
C 10 10 10 20
S  9 10 15 18
S 15 15 15 15
A 12 12 12 16
```

Sample Output

```
Line 1: C 10 10 10 20
        11000 SUM OF CUBES
Line 2: S 9 10 15 18
        INVALID DATA
Line 3: S 15 15 15 15
        900 SUM OF SQUARES
Line 4: A 12 12 12 16
        172 AVERAGE OF THE SQUARES
```

4
Useful Algorithmic Modules

> Pedantry and mastery are opposite attitudes toward rules.
>
> To apply a rule to the letter, rigidly, unquestioningly, in cases where it fits and in cases where it does not fit, is pedantry. Some pedants are poor fools; they never did understand the rule which they apply so conscientiously and so indiscriminately. Some pedants are quite successful; they understood their rule, at least in the beginning (before they became pedants), and chose a good one that fits in many cases and fails only occasionally.
>
> To apply a rule with natural ease, with judgment, noticing the cases where it fits, and without ever letting the words of the rule obscure the purpose of the action or the opportunities of the situation, is mastery.
> G. Polya: *How to Solve It*, p. 148.

Now that we have covered the rudiments of problem solving by using algorithms, we must discover what we can do with what we know thus far. (The last chapter introduced fundamental notions about algorithms and problem solving.)

In problem solving, we discussed the need to be aware of its three major aspects: initial condition or state, required result or outcome, and rules that govern the allowable operations (or ways) of transforming the initial state into the result. We also discussed heuristics we can use to get us thinking along productive lines. (Remember the four steps in problem solving: understanding the problem, devising a plan, carrying out the plan, and looking back. It might be helpful to reread those steps at this point.)

In algorithms, we discussed the ways in which we are allowed to specify data and operations on the data. We introduced variables (named memory cells), assignment of a value to a variable, read and write operations, and control structures that allow us to manipulate the order in which steps in an algorithm are executed.

Now that the fundamentals are understood, we need to inquire further. Two of the most important questions in step 2 of problem solving are "Do you know the solution to a similar problem that you could use?" and "Can you think of a new rule to help you solve the present problem? Can you restate the rule in terms of the other, predefined operations?"

In this chapter, we will "discover" a set of useful algorithmic pieces that appear over and over as parts of many problems. In a sense, we will invent new operations and define them in terms of the old operations. These pieces are also the basis for a "bag of tools" you'll continually use when you seek solutions to similar problems.

4.1. READING DATA

Almost all problems involve reading, using, accessing, and receiving data; and at this stage, almost all problems involve reading in lists of numbers and computing something about them (e.g., the average of a list, its minimum or maximum, etc.). As a result, we should be acquainted with the common ways data lists are presented to you, the problem solver, and for each variation, an algorithm fragment that reads in a list arranged in that variation.

In Chapter 3 we encountered the end-of-file data list in solving the best-MPG problem. To refresh your memory, the algorithm is repeated below.

```
Start algorithm
  write, "The data is:"
  write, "Miles Gallons MPG"
  set BestMPG to 0.0
  while there are data left do
    read, gallons, miles
    if (gallons > 0.0) and (miles > 0.0)
      then
```

```
            set NewMPG to miles/gallons
            write, miles, gallons, NewMPG
            if NewMPG > BestMPG
               then
                  set BestMPG to NewMPG
            end if
         else
            write, miles, gallons, "--- Data must be positive."
            write, " These data ignored."
      end if
   end while
   if BestMPG > 0.0
      then
         write, "The best MPG is: ", BestMPG
      else
         write, "There were no data to be read."
   end if
End algorithm
```

ALGORITHM EXAMPLE 4–1

In an abstract sense, this algorithm has these major pieces:

```
[Initialization section]
While there are data left do the following
   read a datum

      [process the datum just read]

end while loop

[Summary processing involving results]
[obtained from processing the complete list]
```

ALGORITHM EXAMPLE 4–2

In fact, almost any algorithm that processes data arranged in this sort of list has this structure: an initialization section, a while loop which stops after all the data has been read with a read as the first thing in the loop body, and a summary section. At this stage the question arises: "What other kind of data list can there be?"

The second way (of three) of listing data is a list ended by a *sentinel* value. A **sentinel** value is a value that stands at the end of the list and is defined as the *end* of the list. In this sense, we reserve that value as the "end" and so preclude its use as a data value. Consequently, we see lists described as a list of real numbers ended by 9999.0 or a list of positive integers ended by a negative integer, etc. In each case, the sentinel value, or range of values, cannot occur as data, since by definition it signals the end of the list.

For example, given a list of integers, ended by a negative number,

```
 2
 9
13
 5
 1
-2
```

there are five data values. The sixth number, which is negative, indicates there are no more values. In this list,

```
 2
 9
 3
-1
 5
 6
-2
```

we have only three data values. By definition the list ends with the -1, regardless of what follows it.

This technique is commonly used in two applications. First, when there's no way of detecting the end of data on a file. (It is sometimes difficult to find the end-of-file on a terminal,[1] so we resort to the sentinel value to indicate the end.) Second, when we need more than one data list on the same file—a sort of list within a list. For instance, we might want to be able to read two lists, each ended by a negative number (the first list consists of 2, 9, and 3 and the second list of 5 and 6. If we wrote:

```
2
9
3
5
6
```

[1]Different computer programming languages handle end-of-file detection differently. Thus writing an algorithm to find the end-of-file in a uniform manner, regardless of programming language, is sometimes difficult.

we would have no way of determining where the first list ends and the second begins, and so any algorithm we write would read the whole as one list. Now, however, we must write an algorithm to read a list that is ended by a sentinel value.

At first thought (inasmuch as we don't want to reinvent the wheel), we could modify the while loop of the previous algorithm so the condition in the while tests for the sentinel value instead of the end-of-file. If we have a list of positive numbers, ended by a negative number, we could restate the algorithm like this:

```
[Initialization section]
while x ≥ 0 do the following
  read x
  [process x, the value just read]
end while loop
[Summary processing]
```

ALGORITHM EXAMPLE 4–3

But this will not suffice, because—after we complete the initialization section—we check the condition $x \geq 0$ before we enter the loop. At that point, x has been given no known value. This is the problem we encountered when we had to set the sum in the averaging algorithm to zero before we could use it.

A variable that has not been assigned a value has an unknown value (left by a prior algorithm in the memory cell associated with the variable's name). We can't predict the result of comparing x with zero since we don't know what value x contains. To use x, then, would violate one of the principal concepts behind algorithm development: An algorithm must always arrive at the same result, given the same data. If we can't predict what's in x, we can't be sure of always getting the same result given the same data.

To alleviate the problem, we can give x a value, before we enter the loop, by reading a value into x just before the while:

```
[Initialization section]
read x
while x ≥ 0 do
  read x
  [process x]
end while
[Summary processing]
```

ALGORITHM EXAMPLE 4–4

Now x has a value: the first value in the data list. However, this won't work either (for another reason).

As you trace Algorithm 4-4, the first value that's read from the list is compared to zero as we enter the while loop; then another value (the second in the list) is read into x. This means that the first value was ignored: we didn't execute [process x] when x still had the first value.

In frustration, let's remove the second (offending) read, and trace the algorithm. Once inside the loop, no more data is read and, presumably, the value of x remains unchanged. This leads to an **infinite loop** which means the loop is executed over and over, forever. This is not desirable. Unless we exit the loop, we will never write out the results. Furthermore, this could be very expensive, if we run the algorithm on a computer that charges for the time we use.

Having exhausted almost all possibilities, let's put a second read in the loop, but this time after [process x]:

```
[Initialization section]
read x
while x ≥ 0 do
  [process x]
  read x
end while
[Summary processing]
```

ALGORITHM EXAMPLE 4–5

Again, trace the algorithm. This time, for a list of 1, 2, −1, we get

```
[initialize]
read x (x = 1)
while-check: x ≥ 0 is true so execute the loop body
[process x]
read x (x = 2)
while-check: x ≥ 0 is true so execute the loop body
[process x]
read x (x = −1)
while-check: x ≥ 0 is false so exit the loop
[summary processing]
```

ALGORITHM EXAMPLE 4–6

This is precisely the sequence of operations we want to perform. Finally, we've arrived at the standard form of a loop to read in a list that ends with a sentinel value. It is distinguished from the loop to read in a list that ends with an end-of-file, by the fact that it has two reads: one immediately before the while loop and another as the last instruction in the loop body.

The third standard way of listing data is called the **header value list**, in which the first datum, the header, is the number of subsequent items. For example, in

```
 4
 5
 9
-1
 7
```

the first number, 4, indicates that four values follow it: 5, 9, −1, and 7. Of course, as with the sentinel value list, there may be more data after the 7, but we stop there by definition. (There could be another list after the first, or something like that.)

A header value data list is less commonly used than the other two because it is more subject to error: if the person who enters the data miscounts, or if data is added or later deleted, the first value could be wrong and lead to an incorrect result.

This method of listing data is useful, however, when we can neither detect the end-of-file nor reserve a value as the sentinel value—if for example, we need to type the data in at the terminal and the data could be any number. Use of the terminal precludes checking for the end-of-file, and the fact that any value can appear as data means we can't reserve a value as the sentinel.

To read in such a data list, we must first read in the header. Also, based on its value, we must read in that many other data values. Since the number of data values is determined before we enter the loop, we can use a for loop:

```
[Initialization section]
read header
for i ← 1 to header do the following
   read x
   [process x]
end for
[Summary processing]
```

ALGORITHM EXAMPLE 4–7

If we trace the loop on the sample data above, we see that the list of operations is

```
[initialize]
read, header
set i to 1
for-check: i ≤ header (is true)
```

```
read x
[process x]
   (increment i is done automatically by the for instruction)
for-check: i ≤ header (is true)
read x
[process x]
   (increment i is done automatically by the for instruction)
for-check: i ≤ header (is true)
read x
[process x]
   (increment i is done automatically by the for instruction)
for-check: i ≤ header (is true)
read x
[process x]
for-check: i ≤ header (is false—loop stops)
[summary processing]
```

ALGORITHM EXAMPLE 4–8

Thus the loop structure is correct.

Now that we know all three loop structures, we can use them as a base on which to build algorithms. Getting started is always hardest—but now we know three starting points. After we choose the right one, we need only specify the initialization section, process data section, and summary section for the problem at hand.

4.2. COUNTING DATA

To this point in our discovery of algorithms and problem solving, we have seen only the simplest problems, dealing with lists of data. Each item in a list has been independent of all the others, and our algorithms, except for the find-the-largest, have not dealt with summarizing or collecting facts about the data list as a whole. For example, the find-the-largest algorithm produced the largest value in a list.

Several other important algorithmic modules are part of every programmer's bag of tricks, and in this section we will use the template modules (of the last section) to read in a list and count its items. In the next section, we will sum the numbers in the list. Later, we will combine the counting and summing algorithms to produce an algorithm to average a list of numbers.

Given any one of the three kinds of data list in the last section, we want to produce an algorithm to count the items. This, at first, seems simple, and it is—if we start on the right foot. One question we should always ask is "How would we do it by hand?" That's easy, we say; "We count 1, 2, 3, . . ." But again, we must restrict ourselves to variables and algorithmic instructions, and it will help if we think of using a calculator.

To count with a calculator, we add 1 to the display for each new number. At the end of the list, having added 1 for each item, we have the count in the display. Or do we? One of the lessons we learned from the combination-lock problem is that every solution must specify an initial state, intermediate states, and a final state. For the initial state in the counting problem, we clear the display so that it starts at zero. The plan now goes:

```
Clear the display (set it to 0)
Add 1 to the display for each data item as we go through the list
At the end of the list, report the count from the display
```

How do we write this algorithmically? We know we have to read in the data list; so let's start with the template for an end-of-file data loop.

```
[Initialization section]
while there are data left do
  read x

  [process x]

end while loop

[Summary processing involving results]
[obtained from processing the complete list]
```

ALGORITHM EXAMPLE 4–9

It would seem appropriate if the [initialization section] contains the set display to zero. This yields:

```
set display to zero
while there are data left do
  read x

  [process x]

end while loop

[Summary processing involving results]
[obtained from processing the complete list]
```

ALGORITHM EXAMPLE 4–10

To add 1 to the display for each item on the list, we put the instruction set display to display + 1 in the section [process x]. This change yields:

```
set display to zero
while there are data left do
   read x

   set display to display + 1

end while loop

[Summary processing involving results]
[obtained from processing the complete list]
```

ALGORITHM EXAMPLE 4–11

Lastly, we write the result by writing out the value of the display. This goes in the [summary processing] section. If we put it before the while loop, we write out the value of the display before we read any data, which is obviously wrong. If we put it inside the loop with the set display to display + 1 instruction, we would execute the write instruction each time through the loop—each time we read a data value. This would produce the answer, but many other "answers" as well. The only place left is after the loop in the [summary processing] section, and the algorithm now looks like

```
set display to zero
while there are data left do
   read x

   set display to display + 1

end while loop

write, "The data list contained ", display, " items."
```

ALGORITHM EXAMPLE 4–12

Trace this algorithm using the data list

```
 5
-7
 6
 0
```

Convince yourself that the output really is "The data list contained 4 items." Review the heuristic questions that are asked at the third stage in problem solving. Are any changes needed?

The algorithm is almost finished. We continue to write the data after we read it. (Only in this way do we know that the algorithm reads the data we think it is reading.) This becomes more important when we translate the algorithm into a program and try to run it on a computer. It especially helps in a phase of programming known as *debugging*. (A program **bug** is an instruction, or lack of an instruction, that causes the program to behave in a way other than we expect. That is, the algorithm/program is wrong because we gave instructions that did not match our intentions.) The final version of the counting algorithm appears below:

```
start algorithm
  set display to zero
  write, "the data list is:"

  while there are data left do
  read, x
  write, x

  set display to display + 1

  end while loop

  write, "the data list contained ", display, " items."
end algorithm
```

ALGORITHM EXAMPLE 4–13

4.3. SUMMING DATA

Once we have a bagful of handy algorithmic modules, we ask one of our most useful questions: "Is this problem similar to one we have already solved?" (from stage 2 of problem solving). Even at this point, the question is useful. The problem is to sum a list of numbers, and the solution can easily be found if we see that summing and counting are essentially the same. The only difference is that instead of adding 1 to the display for each item, we add in the data value we've just read.

To convince ourselves, think again of a calculator. How do we add a list of numbers? Is it similar to the way we use a calculator to count values? Trace the modified algorithm below and see how it works.

```
start algorithm
  set display to zero
  write, "the data list is:"

  while there are data left do
    read, x
    write, x

    set display to display + x

  end while loop
  write, "The data list totals ", display
end algorithm
```

ALGORITHM EXAMPLE 4–14

This is a trivial example, but it shows the importance of structural abstraction in noticing the similarity between problems. Counting and summing both require a single pass through the data list. Both begin by initializing a sum to zero. Both end with a single output of the result, after examining each value in the data list. Both have the same structure: the basic read loop.

4.4. AVERAGING DATA

Now that we have algorithms to sum and to count a list, it's easy to write an algorithm to average a list. An average, after all, is the sum of numbers in a list divided by their count. Could we sum, then count them, by putting the two algorithms back to back? This might seem to work, but if we trace the joined algorithm we run into trouble. After the summing algorithm stops, the data list has been completely read, and the counting algorithm assumes that it, too, must read the data. Unfortunately, by that time no data is left. So we must take a different approach.

If we can't join the two algorithms end to end to get the result, maybe we can join them side to side. Actually, we need to merge them. First, we identify structures that fulfill the same purpose in both algorithms and, separately, those that serve different purposes. The basic end-of-file read loop underlies both summing and counting algorithms. The instructions that make each algorithm unique are those that add and write. From the counting algorithm, we get the unique instructions

```
set display to display + 1
```

and

```
write, "The data list contained ", display, " items."
```

From the summing algorithm, we get

```
set display to display + x
```

and

```
write, "The data list totals ", display
```

Both algorithms have the instruction set display to zero at the beginning. They are, however, different displays: a variable that contains the running sum of the data, and another variable to contain the count. In our conmbined algorithm, we must make these variables different; so we will call them by different names, sum and count (to be obvious).

To combine the two algorithms, we merge the instructions that appear in the [initialization section] of the read loop. Merging the instruction from the counting algorithm into the summing algorithm and changing names, we get

```
start algorithm
  set count to zero
  set sum to zero
  write, "the data list is:"

  while there are data left do
    read, x
    write, x

    set display to display + x

  end while loop

  write, "The data list totals", display
end algorithm
```

ALGORITHM EXAMPLE 4–15

Then we set the instructions in the [process x] section:

```
start algorithm
  set count to zero
  set sum to zero
  write, "the data list is:"

  while there are data left do
    read, x
    write, x
```

```
        set count to count + 1
        set sum to sum + x

    end while loop

    write, "The data list totals", display
end algorithm
```

ALGORITHM EXAMPLE 4–16

The write statements from the two algorithms are still appropriate. It would be nice to know the sum and the count, in addition to the average, but we have yet to compute the average. We do this after we have written out the sum and count.

```
start algorithm
    set count to zero
    set sum to zero
    write, "the data list is:"

    while there are data left do
        read, x
        write, x

        set count to count + 1
        set sum to sum + x
    end while loop

    write, "The data list contained ", count, " items."
    write, "The data list totals ", sum

   set average to sum/count

   write, "The average of the data list is ", average
end algorithm
```

ALGORITHM EXAMPLE 4–17

If we trace this algorithm, we see that it works most of the time; but it has an annoying flaw. What would happen if (for some obscure reason) someone gives the algorithm an empty data list? Everything works fine, until we try to compute the average. The values of the variables at that point are (and should be) sum = 0 and count = 0. The output to that point is

```
the data list is:

The data list contained 0 items
The data list totals 0
```

If we instruct the computer to set average to sum/count, however, we are in trouble. Division by zero is undefined in mathematics and in the computer's electronics. To avoid this, we check the value of count before we compute the average. If count is greater than zero, we can proceed. If count equals zero (the only alternative), we print a message saying we can't average a list of numbers that doesn't exist. Because we must choose between two courses of action, we use an if-then-else:

```
start algorithm
  set count to zero
  set sum to zero
  write, "the data list is:"

  while there are data left do
    read, x
    write, x

    set count to count + 1
    set sum to sum + x

  end while loop
  write, "The data list contained ", count, " items."
  write, "The data list totals ", sum

  if count > 0
    then
      set average to sum/count
      write, "The average of the data list is ", average
    else
      write, "The data list is empty."
      write, "No average can be computed."
  end if
end algorithm
```

ALGORITHM EXAMPLE 4–18

Now our algorithm is as sturdy as we can reasonably make it, and will inform anyone who tries to run it with no data that this cannot be done.

4.5. SUMMARY

This chapter discusses the notion of a module, or a solution to a subproblem. We described the module solution to a number of common subproblems: the ways of reading the three kinds of data lists, counting data, summing data, and (a combination of the previous two) averaging.

The algorithm module to read end-of-file data lists is

```
[Initialization section]
While there are data left do the following
   read a datum

   [process the datum just read]

end while loop

[Summary processing involving results]
[obtained from processing the complete list.]
```

ALGORITHM EXAMPLE 4–19

The algorithm module for reading a sentinel value data list is

```
[Initialization section]
read x
while x ≥ 0 do
   [process x]
   read x
end while
[Summary processing]
```

ALGORITHM EXAMPLE 4–20

The algorithm module for the header value data list is

```
[Initialization section]
read header
for i ← 1 to header do the following
   read x
   [process x]
end for
[Summary processing]
```

ALGORITHM EXAMPLE 4–21

The algorithm for counting a data list (of the end-of-file type) is

```
Start algorithm
  set display to zero
  write, "The data list is:"

  while there are data left do
    read, x
    write, x

    set display to display + 1

  end while loop

  write, "The data list contained ", display, " items."
end algorithm
```

ALGORITHM EXAMPLE 4–22

The algorithm for summing a data list (of the end-of-file type) is

```
Start algorithm
  set sum to zero
  write, "The data list is:"

  while there are data left do
    read, x
    write, x

    set sum to sum + x

  end while loop

  write, "The data list totals ", sum
end algorithm
```

ALGORITHM EXAMPLE 4–23

The algorithm for averaging a data list (of the end-of-file type) is

```
start algorithm
  set count to zero
  set sum to zero
  write, "The data list is:"
```

```
    while there are data left do
      read, x
      write, x
      set count to count + 1
      set sum to sum + x

    end while loop

    write, "The data list contained ", count, " items."
    write, "The data list totals ", sum

    if count > 0
      then
        set average to sum/count
        write, "The average of the data list is ", average
      else
        write, "The data list is empty."
        write, "No average can be computed."
    end if
  end algorithm
```

ALGORITHM EXAMPLE 4–24

4.6. EXERCISES

4.01. Rewrite the counting algorithm, assuming the data list consists of positive (nonnegative) integers ended by -1.

4.02. Why is the counting problem trivial for a header value data list? Write the algorithm anyway.

4.03. Rewrite the summing algorithm, assuming the data list consists of positive (nonnegative) integers ended by -1.

4.04. Rewrite the summing algorithm to use a header value data set.

4.05. Rewrite the averaging algorithm, assuming the data list consists of positive (nonnegative) integers ended by -1.

4.06. Rewrite the averaging algorithm to use a header value data set.

4.07. Pick one of the problems at the end of Chapter 3, and assume it specifies a sentinel value data list instead of an end-of-file data list. Describe the changes you'd make to its sample data list and write an algorithm to solve the problem.

4.08. Pick one of the problems at the end of Chapter 3, and assume it specifies a header value data list instead of an end-of-file data list. Describe the changes you'd make to its sample data list and write an algorithm to solve the problem.

4.09. Write a program to

1. Read input (positive integers) from a data file
2. Find if the number is odd or even
3. Print: <number> "is odd"
   or: <number> "is even"
4. Multiply all the odd numbers together and all the even numbers together
5. Print: "The product of all the odd numbers is" . . .
   "The product of all the even numbers is" . . .

Each line of data file contains 1 positive integer, and the first number represents the number of numbers in the file. Consider what the algorithm should do if the data is all even or all odd.

Sample Input

```
4
1
2
3
5
```

Sample Output

```
There are 4 numbers.
  1 is odd
  2 is even
  3 is odd
  5 is odd
The product of all the odd numbers is 15
The product of all the even numbers is 2
```

4.10. The data file contains a list of integers, ending with a zero. Find the sum of all the positive numbers, the sum of all the negative numbers, the average of the positive numbers, and the average of the negative numbers.

Sample Input

```
-4
 1
-2
-5
 3
 0
```

Sample Output

```
-4 is negative
 1 is positive
-2 is negative
-5 is negative
 3 is positive

The sum of the positive numbers is    4
The sum of the negative numbers is  -11
The positive average is   2.0000
The negative average is  -3.6666
```

4.11. Write an algorithm that tabulates and prints a grocery bill for a customer, then calculates the customer's change. Each data line contains a 1-character item code and the purchased item's price in pennies. The last line contains the amount handed to the cashier by the customer, with a code of '$'.

Print a grocery bill, with the amount paid by the customer, then calculate the number of dollars, quarters, dimes, nickels, and pennies the customer should receive in change. Print this result in a suitable format. (Print out only the actual change the customer should receive; i.e., do not print "0 Nickels" if the customer is not supposed to receive any nickels.)

Sample Input

```
S  125
C  549
$ 1000
```

Sample Output

Item	Price
S	125
C	549
Total:	674
Paid:	1000
Change:	326

3 Dollar(s)
1 Quarter(s)
1 Penny(ies)

(You can use remaindering to calculate the change. If Change is an integer variable, Change ÷ 100 is the number of dollars to return, and Change – Dollars*100 is the remainder. This is the technique we used to break the rightmost digits from a number in Algorithm Example 3–20.)

4.12. A golf ball is dropped from various heights (measured in feet) onto a concrete floor or a rug. When the ball drops on a concrete floor, each bounce returns it to 85% of the height of the previous bounce. When the ball is dropped on a rug, it bounces to 20% of the previous height. Write a program that counts, and prints out, the number of bounces the ball makes until the rebound height is less than 1 foot. Your program should also say what surface the ball was dropped on. The data file has one height per line. Each height is followed by 'C' (for concrete) or 'R' (for rug). The data file ends with a negative height.

Sample Input

```
15 C
 2 R
30 R
-3 C
```

Sample Output

On a concrete floor from 15 feet the ball will bounce 16 times

On a rug from 2 feet the ball will not bounce

On a rug from 30 feet the ball will bounce 2 times

4.13. The famous numerologist, Dr. I. J. Matrix, has hired you to help him search for mystical numbers (a number whose sum of digits is equal to their product). E.g., 123 is a mystical number, because 1+2+3 = 1*2*3 = 6.

Dr. Matrix has a data file of numbers he thinks might be mystical, and you are to write a program that will compute the sum and the product of digits in each number in the data file. Print out whether or not each number is mystical. Your program should stop when it reaches a negative number on the data file.

Your program should work regardless of how many digits are in each number. To get the rightmost digits, take the remainder after you divide by 10.

Sample Input

```
   22
 1234
13131
   -1
```

Sample Output

```
The number 22 is a mystical number
The sum of the digits is                    4
The product of the digits is                4

The number 1234 is not a mystical number
The sum of the digits is                   10
The product of the digits is               24

The number 13131 is a mystical number
The sum of the digits is                    9
The product of the digits is                9
```

5
Computers and Algorithms

> All programmers are optimists. Perhaps this modern sorcery especially attracts those who believe in happy endings and fairy godmothers.
> Frederick P. Brooks, Jr., *The Mythical Man-Month*

5.1. THE COMPUTER AS A MACHINE

Now, as we consider various properties of the machine that executes our program, we've reached the point at which we must understand the machine in terms of programming language. Indeed, our generalized concept of algorithmic structures and problem-solving includes tools and operations, so that when a solution is implemented on a computer, we know that the computer itself becomes the tool. Hence solutions must be stated in terms of the operations the computer can perform. This puts certain constraints on the problem solver, who must define the tool and its set of operations explicitly.

First, we review the concept of *program*, which is a list of instructions in a formal programming language (e.g, FORTRAN 77 or Pascal. Each instruction conforms to the precise rules of the language. A programming language is necessary because it isn't enough to tell a computer what to do in plain English (even if we could). "Natural" languages are not sufficiently precise to specify orders to a computer.

A computer program is a version or realization of an algorithm, and to convert an algorithm to a computer program (called **programming**), we must consider several things. The machine (computer) must have the speed, memory capacity, and accuracy to execute a program correctly and in a reasonable amount of time. In Chapter 3, we saw that a computer has the important components of input-output, processor, and memory. The next section discusses the basics of computers and their attributes and capabilities.

Figure 5–1 represents a typical computer architecture, which is by no means unique; many similar machines are on the market. Moreover, the basic characteristics and operating principles of computers have changed very little over the years. The "computer revolution" has been centered on such secondary characteristics as size, speed, cost, and physical appearance. Recently, several significantly different architectures have been suggested for the so-called "fifth generation."

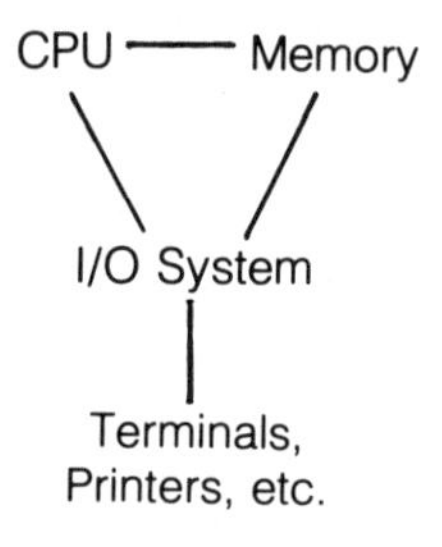

FIGURE 5–1 Typical Computer Architecture

A computer, of course, must be able to receive information. If we write an algorithm to count the number of zeros in a sequence (as in 3, 5, 0, 1, 2, 6, 0, 0, 5, 1), we must have a mechanism for inputting the numbers, and the primary means is a typewriter-like device called a **computer terminal**. A terminal encodes the characters (letters and numbers) electronically and transmits this information to the input function of the computer. The wide variety of other computer input devices includes magnetic tapes, magnetic disks, optical scanners, television cameras, magnetic ink scanners, and so forth. Speed, cost, and ease of use are important in an input device. However, any device that is capable of producing numeric or alphabetic information can be connected to a computer as an input device.

Next is the **input/output** (I/O) system, or the hardware that's required to move program information between "memory" and the outside world or vice versa (e.g., to "read" the characters typed at a terminal keyboard into memory.) Hence the I/O system is an interface between the central processing unit, memory, and the outside world.

The **memory** of a computer is a collection of cells each of which holds a character, integer number, real number, or logical constant ("true" or "false"). Such a cell is usually called a **word** or **byte**. Memory has the capacity to retain and recall information. (Many hand-held calculators also have this feature, but only enough memory to add, subtract, multiply, and divide two numbers and produce the results.) The large memory capability ("size") of a computer distinguishes it from a simple calculator. (In the past few years, the cost of producing large memories has been falling rapidly; hence very large memories are becoming common.) Most computer systems have a variety of types and speeds of memory.

At the heart of the computer is the **Central Processing Unit** (CPU), the most complex part of a computer system. It takes data from memory, processes it, then returns it to memory. The CPU can perform addition, subtraction, multiplication, and division. It can also compare numbers and characters, then make decisions concerning its next operation based on the outcome of such comparisons. It can also perform very primitive input and output operations.

The most fundamental principle in the CPU is automatic operation: the ability to operate with very little human direction or assistance. This principle, embedded in every aspect of computer design, has the effect of increasing both the speed and accuracy of the system. (This notion will be expanded when we discuss the concept of programming.)

The computer, finally, must communicate its answers to the initiator of a problem, and this transmission of information in a form external to the computer is called **output**. Programs usually "write output", which means they print the result on a **cathode ray tube** (CRT), plotted or encoded on a magnetic tape or disk. In modern computer systems, the principal form of output is the printed page (the easiest form to use).

As we have seen, any function that can be computed by following an algorithm can be computed by a digital computer, which executes a corresponding program.

Computation by hand or with a simple hand-held calculator requires human direction to initiate all basic operations (though they are completed automatically), whereas a very important feature of a computer system is *programmabil-*

ity. A computer is **programmable** if a meaningful sequence of basic operations (addition, subtraction, etc.) can be selected in advance and the entire operation can be executed automatically. This preselected sequence is stored in memory; so it can easily be repeated or rerun. (Programmability distinguishes a hand-held calculator from a computer.) The only limitation on the number of preselected sequences is the amount of memory (because the needed amount may exceed the amount available).

A practical consideration is the *speed* of the central processing unit—a solution may require so many steps that it cannot be done in a reasonable amount of time. The classic examples are game playing (chess, GO, and so on), because a computer does not have the capacity to store all possible moves. In chess, for example, there are 10^{120} moves—far beyond the capacity of all computers. In a 40-move line of play, the number of possibilities is beyond the capacity of any of today's digital computers. (Techniques that can solve such problems are in an area of computer science called *artificial intelligence*.) Some problems do not require huge memories, but take more than a reasonable amount of time (we'll see such a problem—the travelling salesman—in the next section).

Another aspect is the wide range of computer systems. To compute a few planetary positions might take 15 minutes on an Apple II but only a few milliseconds on a CRAY-I. A CRAY-I is an example of a *super-computer*.

An important consideration is *accuracy*. Frequently, in ordinary calculations, we use values that cannot be expressed in a finite number of digits (e.g., the square root of 2 = 1.414 . . . and pi = 3.1415926535 . . .). Since a memory cell can store only a finite number of digits, we compromise: we adopt the convention that every number used in computing is obtained by taking a finite number of digits (from its infinite digit representation) to *approximate* the number.

In this sense, all numbers are approximations and may be written as M = N + E, where M is the actual number, N is the approximate number, and E is the error. When M can be represented exactly by the finite-digit approximate number, N, E is zero. (In particular, integer arithmetic in most computers deals with exact numbers [integers] and produces exact results [integers].) E may be positive or negative, but is generally assumed to be small compared to N. The absolute error in N is defined as the absolute value of E.

As an example of accuracy, consider a computer that has a maximum memory cell size of 8 digits—the largest number it can store consists of 8 digits. When the number 0.10000000 is multiplied by 10.0, the answer is hardly ever 1.0000000; it is usually 0.99999999. Hence the value of E is 0.00000001. A more complicated example is the numbers 123.45677 and 123.45678. The number 123.45678 may be as large as 123.456785 and the number 123.45677 may be as small as 123.456765. The computed difference between them is 0.00001; the actual difference is between 0 and .00002. Hence a computed difference may have an error as great as 100 per cent. Accuracy, therefore, is important in every computer system.

5.2. FEASIBLE AND INFEASIBLE COMPUTATIONS

In the first four chapters, we developed a notation for describing and communicating algorithms that is natural and easy to understand. It consists of five very basic algorithmic structures:

assignment
if-then-else
read-write
while-do
for

To see how these structures are embedded in our design of an algorithm, consider this example.

```
1. Set SUM and COUNT to 0
2. while there is data left do
3.     read the next number off the data list, store it in X
4.     add X to SUM
5.     add 1 to COUNT
6. end while
7. compute SUM/COUNT, and store the result in AVG
8. write "The average of", COUNT, "data values is", AVG
9. stop
```

ALGORITHM EXAMPLE 5–1

In terms of our algorithmic structure, step 1 is an assignment. Step 2 is a while do. Step 3 reads information. Steps 4 and 5 are expressions and assignment. Step 7 is a combination of expressions (SUM/COUNT) and assignment (store the result in AVG). (An obvious improvement would be a check to see that COUNT is equal to zero.) Step 8 writes out information. (As we saw in the previous section, the answer is output in our language.) Step 9, although it is obvious, is important and necessary: we must be able to stop a computer algorithm.

The approach in this book is *hierarchical algorithm design*, which means that a program is constructed by focusing on a problem and developing a series of algorithms, starting with a general one and refining it into specifics, each representing a correct solution to the problem. This development leads us to the final algorithm, which uses the five algorithmic structures. The final step is to translate the algorithm into a program, using a programming language. The algorithms are retained for *human* readability whereas the computer program is *machine* readable. People can read and execute algorithms and programs, but the machines can only read and execute programs.

These examples can be transformed into an algorithm that uses the five basic algorithmic structures:

```
Start Algorithm
  set SUM to 0
  set COUNT to 0
  while there is data left do
     read X
     add X to SUM
     add 1 to COUNT
  end while
  set AVG to SUM/COUNT
  Write, "The average of:", COUNT,
     " data values is: ", AVG
End Algorithm
```

ALGORITHM EXAMPLE 5–2

When an algorithm is in this form (Example 5-2), it is easy to rewrite it in a programming language. To do so, we must consider the capabilities of our machine: Does it have the necessary accuracy, speed, and memory to execute the program correctly and in a reasonable amount of time?

You may find this surprising, but a computer cannot solve *every* problem. How, then, can we address its drawbacks? Most often we are interested in the computer time or memory space required to solve a large problem, and we measure the quantity of input data (or *size* of a problem) by comparing alternate algorithms. In a previous example, the size of the problem was the number of items we read from the data list (step 2). If N denotes the size of the list of data we used in the averaging problem, the time this algorithm needs is the number of time units required to process an input list of size (length) N. For example, if x is the time required to process 1 data item, Nx is the time required to process N items.

N	Time
1	x
10	10x
100	100x

Hence the time required to process 100 data items is 100x. The space needed by this algorithm is the number of memory units required to process an input list of length N.

Time and memory-space size are important in solving a problem on a computer (which has limited resources), and a problem may require too much memory or time, or both. Hence the rate of growth of time or space is an important measure of the efficiency of an algorithm.

Or consider the usefulness of manipulating a table of numbers or matrix—the 5 × 5 matrix in Figure 5-2a. Each space in the matrix can be thought of as containing a number. For example, the number in the square in row 3 and column 4, denoted by M(3,4), is 14; and, in general, the number in the square in row i and column j is denoted by M(i,j).

columns

rows	1	2	3	4	5
1	1	2	3	4	5
2	6	7	8	9	10
3	11	12	13	14	15
4	16	17	18	19	20
5	21	22	23	24	25

(a) Matrix

	1	2	3	4	5
1	1	6	11	16	21
2	2	7	12	17	22
3	3	8	13	18	23
4	4	9	14	19	24
5	5	10	15	20	25

(b) Matrix Transpose

FIGURE 5–2 A Matrix and Its Tranpose

Interchanging rows and columns of a matrix (transposition), moreover, is important for many applications, because transposition is equivalent (in this instance) to replacing M(i,j) by M(j,i) and M(j,i) by M(i,j)—as depicted in Figure 5-2b.

It is useful to specify the order in which we transpose a 5 × 5 matrix. The values of i and j, which must be considered when the matrix is transposed in the order determined by Figure 5-3a, are depicted in Figure 5-3b.

	1	2	3	4	5
1		1	2	3	4
2			5	6	7
3				8	9
4					10
5					

(a)

I	*J*
1	2,3,4,5
2	3,4,5
3	4,5
4	5

(b)

FIGURE 5–3 Carrying Out a Transposition

```
Start Algorithm
  read N and the matrix M
  set I to 1
    while I is less than or equal to N-1
      set J to I+1
      while J is less than or equal to N
        set T to M(I,J)
        set M(I,J) to M(J,I)
        set M(J,I) to T
        set J to J+1
      end while
      set I to I+1
    end while
  write, the matrix M
End Algorithm
```

ALGORITHM EXAMPLE 5–3

The result of inputting N = 5 (the number of rows or columns) and the matrix M, the transposition in Figure 5-2a is given in Figure 5-2b. Figure 5-3a shows the number of times the exchange in the inner while-do must be executed. The number of interchanges was $N(N-1)/2 = 10$. In this example, the number of memory cells required to process an input of size N is N^2, the size of the matrix. For large N, this could be a major problem in the computer solution.

The time needed by this algorithm is the number of time units required to process an input list of size N. If x is the time required to process a list of size $N = 1$, N^2x is the approximate time required to process a list of size N (i.e., $N(N-1)/2$). For example:

N	Time
1	x
5	25x
10	100x
15	225x

FIGURE 5-4 Data List Length and Computing Time

Hence as N increases, the time increases quadratically (i.e., by a power of 2). For $N = 15$, the time required to transpose the matrix is approximately 225x. This algorithm's time requirement increases at a faster rate than that of the simple averaging algorithm.

The traveling salesman problem is a "classical" mathematical challenge to find an optimal route between a series of locations under the condition that each location is visited only once and, the salesman returns to his point of origin. (By "optimal route," we mean that the sum of the distances [or travel cost or time] is a minimum). This problem, then, involves sequencing locations to minimize a function of the travel between them. It can be represented as a network, where the nodes represent the towns (locations) and the lines the distances (travel time) between them.

In this four-town problem, the salesman's round trip is given by the lines (1,2), (2,4), (4,3), (3,1), which, taken in order, represent the total trip. (E.g., 1,2

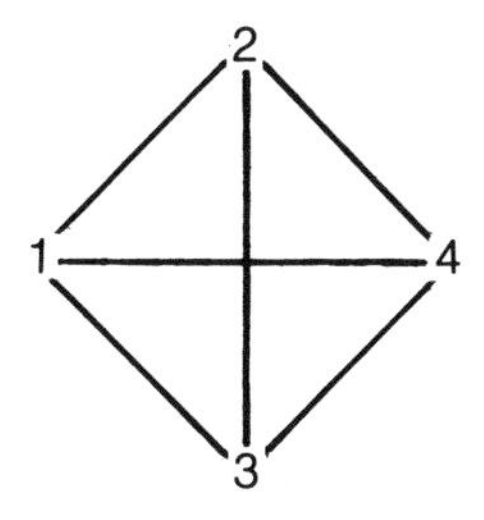

FIGURE 5-5 A Network of Four Towns

represents the travel from node 1 to node 2.) The salesman wants to start from a given town, visit each town once, then return to his starting point; the objective is to minimize his total traveling time.

Clearly, starting from a given town, the salesman will make a total of 3! = 6 possible trips. If there are n towns, the salesman faces a total of (n − 1)! possible trips. Since he goes through all the towns, the optimal solution must be independent of the selection of his starting point. In general, the problem of a salesman traveling through n towns can be extended to n + 1 towns and m salesmen, where $(n!)^m$ sequences must be considered.

One solution is a program that tries all the possibilities, but this may require more time than is available on a computer. For example, if a salesman travels through 21 towns, there may be 20! = 2,432,902,008,176,640,000 sequences. If a computer executes 1 sequence per microsecond (1/1,000,000 of a second) and works 8 hours a day, 365 days a year, it would take almost a quarter of a million years to find a solution.

As we said, it is still surprising to find problems that cannot be solved on a modern computer. Each time we get more processing power, we attempt to solve problems that previously were unapproachable. Through careful analysis of each algorithm, we determine whether these problems can be solved by using the newest computing power. (Appendix A provides more detail on this complex area of algorithm analysis.)

5.3. STRUCTURE OF A COMPUTER PROGRAM

To repeat, a computer program is a problem solution described in a manner that is understandable by a computer. It is a complete set of instructions in a programming language. As pointed out earlier, algorithmic development organizes a problem into modules in such a way that each module's function is simple and easily understood.

Again, a program or algorithm can be stated as a sequence of general steps, each of which, in turn, is described in more detail by substeps. Eventually, successively refining each step, we arrive at an algorithmic module we know how to write or at one of the five basic algorithmic structures. Ultimately, the program is described in simple and easy-to-program components, each of which is called a *procedure* or *module*. The examples we have seen thus far are fairly easy programs and do not require much algorithmic development—except the chess and traveling salesman problems, which are very complex. (The latter require a great deal of algorithmic development, and Chapters 7, 8, and 9 address this type of problem.)

If a program consists of more than one procedure, one procedure must control the others; accordingly, it's called the *main* procedure. The other procedures have various names, depending on the programming technique and language: *subprograms*, *subroutines*, or *functions*.

Many subprograms are built into a programming language, and are called *library* or *intrinsic* procedures and functions. Examples include the trigonometric, logarithmic, and exponential functions. The concept of library procedures is very important since it deals with a program's function and not with the algorithm we use to perform the function. This implies that many algorithms can be used to perform a function. What algorithm is "best" will be discussed in a later chapter.

A computer cannot understand a problem solution that's described in any language other than its own. Therefore, a program must be written in a machine language that enables the computer to understand *nonmachine* language, the program in Pascal, FORTRAN, etc. Such programs (called **compilers** or **interpreters**) have been written for many languages. A computer executes a program in one of these languages indirectly, using a compiler or interpreter.

A programming language procedure consists of statements that are normally of three types. The first, an **executable statement**, specifies some process or action by the computer, and causes the compiler to generate machine-language instructions to carry out the operations in the specified process. (Executable statements include the five basic operations specified earlier, and the algorithm that computed the average of a list of numbers is an example of executable statements.)

The second type of statement, a **declaration**, provides information to the compiler. Declarations cause memory to be reserved, provide information about the mode of operands in arithmetic operations, generate information for use by subprograms, and so forth. A declaration does not cause action during execution of a procedure; it provides information to the compiler. (This is discussed in Chapters 6 and 8.)

The third type of statement, a comment, is stated in English for the sole benefit of those who want to read a procedure. For example, comments may explain how a subprogram accomplishes something—that is, the idea behind an algorithm. A comment is not used by the compiler and is not executed by the computer. After you start programming in a programming language, these three types of statements will become more and more clear.

In short, a program directs the actions of a computer (machine). Statements must be executed sequentially since each statement specifies a process or action for the computer. Thus a computer is useless without a program.

5.4. LIBRARIES OF COMPUTER PROGRAMS

Two concepts are important for constructing good programs. We make a computer solution easier through modularity and algorithm development. This, of course, makes programming simpler, and our algorithms contain fewer errors. (A module is an algorithm that performs a single task, that solves a single conceptual problem [e.g., finds a square root or the largest number in a list].) Since a complex algorithm is a collection of simple tasks, we can solve a complex problem by tackling these tasks one at a time. In other words, if a task is very complicated, it can be divided into subtasks until it is easy to solve. (Moreover, it is sometimes impractical to increase the generality of an algorithm, because to do so would either make it too complex or vastly increase the number of steps to perform the tasks for which it was designed.)

The second important concept is efficiency. As was pointed out, one algorithm is more efficient than another if its cost of execution is less. In general, the cost of a program is directly related to the amount of storage and computer time. The more storage use, the higher the cost; the more computer-time use, the higher the cost. Our objective for an efficient algorithm is minimum storage and minimum time, which is easy to say but difficult to do. In many problems, it is possible to achieve only one of the objectives. Hence a programmer makes trade-offs in developing an algorithm.

Libraries of algorithms are continually developed to attain both of these objectives (so a programmer does not have to reinvent the wheel). Every programming language has a large number of computer programs (library) that solve many simple (and sometimes complex) problems. Because large statistical and mathematical libraries are available in most computer systems, users don't have to write new, complex modules, but can use modules written by someone else for the same purpose. Refining an algorithm into small, self-contained units makes it much easier to use library algorithms to write a computer program.

5.5. KEY WORD LIST

algorithmic development
algorithmic structures
assignment
central processing unit (CPU)
comment
control
declaration
efficient
executable statement
expressions
function
if-then-else
input
machine language

memory
module
output
program
programming
programming language
read-write
sequential
size
space
subprogram
terminal
time
while-do

5.6. EXERCISES

5.01. Draw a diagram like Figure 5-1 for your computer system. In particular, find out what peripheral and secondary storage devices are a part of your computer system.

5.02. Find out the capacity of disk storage on your computer system.

5.03. List the main memory size and main memory access time for your computer system. The memory access time is the time required to transfer one unit (memory cell) of information between the main memory unit and the CPU.

5.04. Find out how integers, floating point numbers, and alphabetical characters are represented on your computer.

5.05. Write a program on your computer that discovers the largest and smallest representable integer.

5.06. List the various programming languages that are available on your computer system.

5.07. Discover whether your computer system supports time-sharing and batch processing. Learn how to sign onto and use both types of processing.

5.08. Discover how to use the text editor on your computer system. Learn how to use some of its simpler features.

6
Fundamental Data Types

> . . . it gradually came to light that all those matters only were referred to Mathematics in which order and measurement are investigated, and that it makes no difference whether it be in numbers, figures, stars, sounds or any other object that the question of measurement arises.
>
> Descartes: *OEuvres*, vol. X, p. 378; discussion of Rule IV in "Rules for the Direction of the Mind"

6.1. WHAT IS A "DATA TYPE"?

Consider the data values 2, 2.0, and '2' and how they differ from one another. In one sense, in our usual interpretation of these symbols, we would say they are different ways of writing the same thing: the number 2. But this is not quite true. If we allow different ways of writing the number 2, there should be a reason for this. What do we mean by the different representations?

The representation 2 is meant to be an **integer**—that is, a whole number, one that can have no fractional part. The representation 2.0 is meant to be a **real**—that is, a number consisting of a whole number and a fractional part. It just happens (in the case of 2.0) that the fractional part is zero. We usually agree that, mathematically, 2 is equal to 2.0. The third representation, '2', is meant to be a character, and not a number at all.

The feature that distinguishes the numbers 2 and 2.0 from the character '2' is the *quotation marks*. In the sense that '2' is not a number but a character like 'A', we cannot say that 2 is equal to '2'; that's mixing apples and oranges. Furthermore, since '2' is not a number, it cannot be used in arithmetic computations: 2 + 2 is 4 and 2.0 + 2.0 is 4.0, but '2' + 2 and 2 + '2' and '2' + 2.0 doesn't make more sense than 'A' + 2 and 2 + 'A', etc.

This may seem to be a subtle and perhaps pointless distinction. We will see, however, that these are precisely the distinctions made by a computer, and so become part of the rules we must play by to program a computer. If a machine makes these distinctions, we must also, so we can use them to our advantage (or ignore them to our peril).

In using various representations of values in algorithms and programs, we must be rigorous and define each representation exactly. Thus each of the three representations for 2 is a different **data type**: **integer**, **real**, and **character**. Most programming languages provide for these three data types, and a fourth named **boolean** or **logical**, consisting of the values *true* and *false*. These four data types are called the **simple data types**.

Now, we must answer the question we asked as the heading of this section: What is a data type? A **data type** is a collection of data values and the definition of one or more operations on those values.

As a formal statement, this definition suffices, but its implications are not obvious. Therefore, we add: the collection of all integer values that are usable on a computer with the operations of assignment, addition (+), subtraction (−), multiplication (*), truncating division (÷), less than comparison (<), greater than comparison (>), etc., is a data type. The collection of all characters in a well-defined character set with the assignment operation is also a data type. The collection {*true*, *false*} with the logical operations "and", "or", "not", and assignment, is also a data type.

Usually all values in a data type share a set of common characteristics. Each integer, i, for example, is a whole number, and has a successor, i + 1, and a predecessor, i − 1 (except that a computer is finite and has a smallest integer with no predecessor and a largest integer with no successor). Each integer differs from its neighbor by 1.

The data type *real*, although numeric, is basically different from integers as far as computers are concerned. In most machines, the value 1 of type *integer* is represented completely differently from the value 1.0 of type *real*. (We can tell a *real* from an *integer* value by the presence or absence of a decimal point and fractional part. Thus 3.0 is a value of type real while 3 is a value of type integer. Numerically, these two values are equal but, we will learn, they are not necessarily interchangeable in algorithms and computer programs. More specific differences will be covered later.)

Generally, data types are used in algorithms and programs to make our intentions more specific and exact. (Indeed, a data type is a formally specified set of values and operations.) They can also be used by programmers to allow the compiler to ensure that they don't accidentally mix two incompatible types (like trying to add the character value '9' to the integer value 3). Each data type is fundamentally different from all others. Their values and operations are incompatible with those of any other. (We shall see the ramifications of this in a later section.)

The most basic data types are **scalar data types**, in which we simply define or enumerate all possible constants of that type. This enumeration is assumed to be ordered, and each constant has one (and only one) of these relationships: greater than, less than, or equal to any other constant of its type. There are four standard scalar data types: integer, real, character, and boolean. To be stored, each value in one of these types requires 1 memory cell.

Scalar data types can be combined to form highly complex **composite data types**. The way the scalars are combined determines the composite types. Furthermore, composite types can be combined to form even more sophisticated arrangements of data, with exceedingly complex interrelationships. Regardless of their complexity, however, all arrangements of data can be viewed as built up from the simple scalar data types. (Composite data types will be introduced and discussed in a later chapter.)

6.2. THE COUNTABLE SCALAR TYPES

6.2.1. Integers

Integer constants are represented only as sequences of digits, with an optional plus or minus sign. Examples of valid integers are

```
27500   (not 27,500)
 -123

    1

    0

  +20
```

(We define an integer in such a way that a comma cannot be used. Because this convention is used in most programming languages, we will use it too.)

The standard operators defined for the integers are

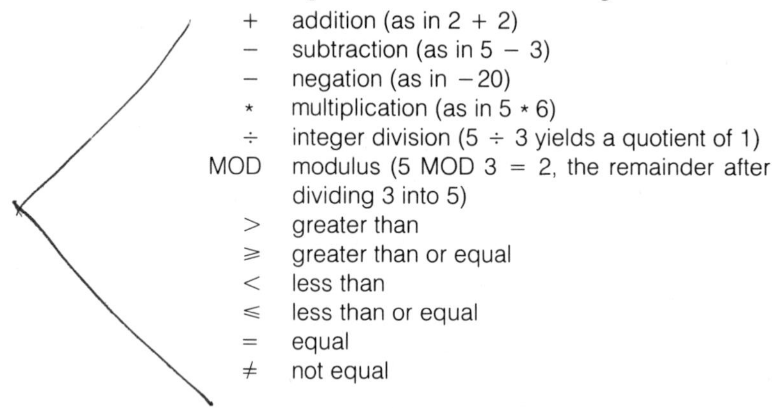

+	addition (as in 2 + 2)
−	subtraction (as in 5 − 3)
−	negation (as in −20)
*	multiplication (as in 5 * 6)
÷	integer division (5 ÷ 3 yields a quotient of 1)
MOD	modulus (5 MOD 3 = 2, the remainder after dividing 3 into 5)
>	greater than
≥	greater than or equal
<	less than
≤	less than or equal
=	equal
≠	not equal

The last six operators, called **relational operators**, operate on any standard scalar data type to produce a result that has the value *true* or *false*.

6.2.2. Characters

Character constants are all the *individual* characters. To indicate an element of the data type *character*, we surround it with two apostrophes or single quotes. To indicate the apostrophe character, we write it twice.

'A'

' ' (a space or the blank character)

'1'

';'

'''' (the apostrophe character)

The elements of the scalar data type *character* are always single characters. Constructs such as 'HELLO' or '***' are not elements of this type, but elements of a more complex data type composed of sequences of characters. (That data type, the **string**, will be described later.)

6.2.3. Boolean

The elements of the scalar type *boolean* are simply the two constants, *true* and *false*, ordered so that *false* < *true*. This may seem strange, but it makes the relational operators applicable and so allows us to make general rules about these operators that apply to all data types.

The relational operators operate on the integer, real, character, or boolean data types and produce a boolean result. In addition, three operators can be applied only to boolean values.

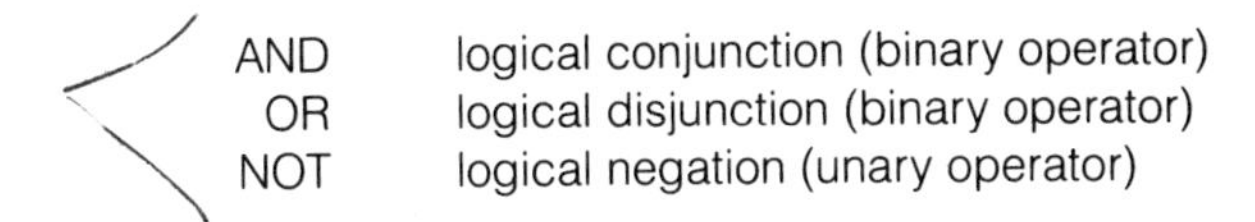

AND	logical conjunction (binary operator)
OR	logical disjunction (binary operator)
NOT	logical negation (unary operator)

These operators are defined by the tables below.

AND	T	F
T	T	F
F	F	F

OR	T	F
T	T	T
F	T	F

NOT	
T	F
F	T

FIGURE 6–1 Boolean Operation Tables

To read the tables, find the table with the appropriate operator and look up the result, using the row and column labels. For example, *false* AND *true* yields *false* according to the AND table, where we find the value in the *false* row and the *true* column. The operator NOT is unary—that is, it takes only one operand—so its table is only one column. The value of NOT *true* is *false*, and the value of NOT *false* is *true*.

6.3. THE UNCOUNTABLE DATA TYPE: REAL

Real constants can be represented in one of two ways: decimal or scientific notation. In decimal notation, a real number is represented by an optional sign, a whole-number part, a decimal point, and a fractional part. In scientific notation, a real number is represented as a value between 1 and 10, the *mantissa*, multiplied by the appropriate power of 10, the *exponent*.

A real constant has a digit on each side of the decimal point (e.g., 0.5 and 5.0) and the following are examples of valid real constants:

```
3.1415927
    -25.0
   +0.198
    5.0E6  (5000000.0)
   -6.0E-8 (-0.00000006)
+6.08E+27
```

We consider the following to be invalid real constants, to follow the conventions in most programming languages:

31	(a valid integer but not a *real* value)
−2.4E7.2	(only whole number exponents allowed)
2,300.5	(no commas allowed)

The operators and functions for the *real* data type are

- \+ addition
- − subtraction (or unary negation)
- * multiplication
- / real division (the fractional part is retained, unlike the integer operator ÷)

In addition, the relational operators discussed earlier are defined for *real* values.

6.4. VARIABLES IN ALGORITHMS

A variable can assume different values during the execution of a program or algorithm. Alternatively, a variable can be thought of as a memory cell inside a computer; it has a name and can hold a value. It is very important to recognize—and remember—the distinction between the *name* associated with a variable and the *current value* of that variable.

X	1.5

FIGURE 6–2 A Named Memory Cell

In Figure 6-2, the name of the variable is X; its current value is 1.5. The name of the variable is permanent and is associated with a memory cell the first time the name appears in an algorithm. It's associated with the same memory cell for the rest of the execution of the algorithm. The value, however, is volatile and may change often during the algorithm's execution. Any of these commands

```
set x to 2.5
add 10 to the current value of x
read in a new value for x
```

leave the name of the variable unchanged but produce a new current value.

Some languages have no restrictions on the values we may assign to a variable. For example, in some languages a variable may have the integer value 6 at one point, then be changed to the real value −2.71. In such languages, usually called

weakly typed or **typeless**, the concept of data type is not fundamental, and there are few or no explicit rules associated with data typing.

We take the opposite approach in which a variable may be given a value from only one type and is therefore called **strongly typed**. This means that

1. Each scalar variable must be explicitly associated with a single scalar data type.
2. The variable may assume values of only that data type for the duration of its existence.
3. Most operators are defined only for specific data types. Applying them to any other data type is treated as an error.
4. Mixing data types (e.g., adding a character variable to an integer constant) is usually an error.

In this chapter, *data type* means the scalar types we have discussed: integer, real, character, and boolean. Later, we will use the term for a larger number of other data types.

In our algorithms, the type of value we first store in a variable determines the data type of that variable. Once we have a data type for a variable, we have automatically defined the range of values it may assume and the set of operations that can be performed on it. Some people like to state the variable names explicitly and the data type associated with each variable at the start of an algorithm. This is called **declaring** the types of the variables and the statements are called **declarations**. Most programming languages *require* declarations for all variables in a program to make the programmer decide upon a type and to provide a list of variables and their types for checking errors and for later reference.

Now we can distinguish between the two classes of operations that are performed on variables.

First, defining a variable is the process of either creating a value for or changing the current value of a variable. When a variable is redefined, the new value replaces the old value, and the old value is lost. All of the following algorithmic commands are valid ways to define an integer variable K.

```
set K to 3
increment K by 1
read in a value for K
```

To attempt to do something such a thing as

```
set K to 123.456
```

is invalid, since K is declared an integer and 123.456 is a real value. We must restrict ourselves to integer values when we define K.

Second, referencing a variable is the process of using the current value of a variable.

write out the current value of K
set L to the current value of K

It should be obvious that a variable must be defined before it can be referenced. In fact, failure to define it leads to an error. It is also the rule, in almost all programming languages, that referencing a variable never changes its current value. In a sense, we can imagine we make a copy of a variable each time we wish to look at or use its value; we do not change the value by copying it. For example, neither of the two algorithmic statements above would cause a change to the current value of the variable K.

6.5. SUMMARY

In this chapter we discussed the notion of *data type* as a collection of values and associated operations. We discussed *constants* and *variables* with regard to data types. In particular, we discussed the four standard scalar data types available in most programming languages: integer, real, character, and boolean. We also defined the operations that normally are available for each of these data types, and alluded to the possibility that these data types can be composed into more complex arrangements than is allowed by a variable of a scalar type.

We also discussed the concepts of strong and weak typing for variables. We take the strong-typing approach as our usual assumption. Finally, we discussed declaration, definition, and reference of variables.

6.6. EXERCISE

6.01. In most programming languages, an integer value may be used wherever the rules require a *real* value for strong-typing. It is not usual, however, to allow a real value wherever an integer value is required. Can you suggest why? (*Hint*: Consider the set of all integers versus the set of all reals.)

6.02. Several expressions involve integer operations. Show the value of each expression below.

5 ÷ 3	3 ÷ 2	6 mod 3
6 mod 4	2 ÷ 4	4 ÷ 2

6.03. All programming languages have the notion of **operator precedence**, which means that some operations (like multiplication and division) are performed before others (like addition and subtraction). For example, 5 + 4 * 2 yields 13. The multiplication, 4 * 2, is done before the addition of 5. This is the way we usually do things in algebra. Evaluate each expression below.

```
5 + 3 * 2 + 4          (5 + 3) * (2 + 4)
(5 + 3) * 2 + 4        5 + 3 * (2 + 4)
7.0 – 6.0/2.0          (5.0/2.5)/2.0
5.0/(2.5/2.0)
```

6.04. Real arithmetic on a computer is not completely accurate (as was shown in Chapter 5), and this can have serious ramifications. Presume—for this exercise—that the computer you use has an accuracy of 6 significant figures or decimal places (when all numbers are represented in scientific notation). Also, you work for a bank and have written a program to compute daily interest of 7.5% on savings accounts. The bank holds over $10 billion in deposits. How much could you steal each day, because of inaccurate real computer arithmetic, without being caught by the auditors? (Assume the average error is 0.5E – 06 times the result of any multiplication.)

7
Modules, Subalgorithms, and Abstraction

. . . divide each difficult problem that you examine into as many parts as you can and as you need to solve them more easily.

Descartes: *OEuvres*, vol. VI, p. 18; "Discours de la Methode," part II

This rule of Descartes is of little use as long as the art of dividing . . . remains unexplained. . . . By dividing his problem into unsuitable parts, the unexperienced problem-solver may increase his difficulty.

Leibnitz: *Philosophische Schriften*, edited by Gerhardt, vol. IV, p. 331

7.1. WHAT IS MODULARITY?

We have discussed several algorithmic modules: how to read each of the three standard ways of arranging data (end-of-file, sentinel, and header), to count the data, to sum the data, to average the data, to find the smallest and the largest value in the data. In a way these modules form a set of building blocks with which we can construct more complex algorithms. Each module implements a concept that we, as humans, find more comfortable to use than the concepts demanded by the machine (assignment, if-then-else, while loops, for loops, read, and write). Using modules that implement higher conceptual operations is easier because we don't have to worry about as many details. This lets us concentrate more on the overall solution.

We digress at this point to discuss why the sequential nature of algorithms plays an important role in our solutions. So far, we have only said that algorithms are sequential; that is, the steps are performed one at a time in the order specified by the algorithm. We can modify the normal sequential flow by using the control instructions: if-then-else, while, and for. But if we look at each of them as a single instruction (even though they may contain other instructions), our algorithm remains sequential: it starts at the first instruction and proceeds to the second, third, and so on through the last, in orderly fashion. Even with the loops and the if-then-else, we perform a sequence of instructions when the algorithm "runs." The important thing is that the instructions be followed in exactly the order specified; otherwise the algorithm might not work correctly. If we shuffled the lines in an algorithm, randomly changing the order we had specified, it probably wouldn't work, and possibly wouldn't be an algorithm.

The order of instructions in an algorithm is important because each instruction takes one or more variables (and possibly the data list or output list), changes them, and transmits them to a later instruction. This next instruction does only what we want, because the previous instructions set everything up. We can see this work in the two-instruction sequence:

```
...
read, x
set sum to sum + x
...
```

For the read, x to work, there must be data on the data list. If there isn't, this instruction causes an error and the algorithm fails. So some previous instruction (an if or a while) must make sure there is data *before* we get to this instruction. This is called a **precondition**—a condition that must be true before the current instruction will do what we want it to. If the read, x works, the result is that the variable x contains the next data value, and the result of executing an instruction is called a **postcondition**. The preconditions for the instruction set sum to sum + x are that the values of the variables sum and x have been preset.

In the summing algorithm, we set the value of sum to zero at the beginning so that the first time set sum to sum x is executed, the precondition is met. The postcondition is that the value of sum has been updated by the addition of x. If the two-instruction sequence (above) appears inside a loop, as it does in the summing algorithm, the postcondition sum of set sum to sum + x becomes a precondition in the next pass through the loop.

The pre- and postconditions are really the values of all the variables we have used at any point in the algorithm and the current state of the data list and output list. Instructions that require no preconditions are all of the type set x to 0, where we set a variable to a constant value. The postcondition of such an instruction is simply that the variable is set to the given value. All other types of instructions require at least one precondition to work properly. For example,

```
while there is data left do
```

requires that someone has created a data list. Even if it was empty to start with, something that can be designated "the data list" had to be made. In the instructions

```
while x > 0 do
```

or

```
if x = 10 then ...
```

or

```
write, x
```

the precondition is that the variable x has an appropriate value.

The postconditions of an instruction are the new values of any variable the instruction changed or the new states of the data list or output list. An instruction that has no postcondition is irrelevant to the algorithm and can be removed (i.e., set x to x accomplishes nothing, has no postcondition, and may safely be removed from the algorithm without affecting any other result).

The postconditions of the instructions of assignment, read, and write are straightforward. Assignment changes the value of some variable; the changed variable is the postcondition. Read changes the state of the data list and the values of the variables listed after the read. Write changes only the state of the output list. These changes are the postconditions. The postconditions of a control instruction such as if-then-else, while loop, or for loop are more difficult to specify exactly, due to their conditional nature.

In general, the postconditions of a control instruction consist of all the postconditions that result from doing the instructions inside the control instruction.

```
if x > 0
  then
    set sum to sum + x
    set count to count + 1
  else
    write, x, " is non-positive."
end if
```

ALGORITHM EXAMPLE 7–1

For example, in Algorithm 7–1 the pre- and postconditions are dependent on the value of x at the time the if instruction is executed. That x must be preset is an absolute precondition since x is compared to 0 every time the if is executed. If x is positive (>0) when the if is executed, then the other preconditions are that sum and count must also be preset and the postconditions are the changed values of sum and count. If x is not positive, then the only precondition is that x is preset, and the postcondition is the changed output list.

We begin to see that an algorithm works only if, at each instruction, the set of postconditions from previous instructions satisfies the preconditions needed by the current instruction. Furthermore, the current postconditions, plus any postconditions added by executing an instruction, must satisfy the preconditions of the next instruction to be executed, and so on, from the first instruction through the last.

It is also important that we can determine a set of pre- and postconditions for blocks of instructions, as in Algorithm 7–2 (to find the largest number in a list).

```
Start algorithm
   read, largest
   while there is data left do
      read, value
      if value > largest
         then
            set largest to value
      end if
   end while
   write, "The largest value in the data list was: ", largest
   stop
End algorithm
```

ALGORITHM EXAMPLE 7–2

The precondition is a data list that contains at least one value (why can it not be empty?). Of the three postconditions, the most visible is that the output list has "The largest value . . ." written to it. The other postconditions are that (1) the variable value contains the value of the last data item and (2) the variable largest contains the value of the largest data item.

Looking at the pre- and postconditions of algorithm segments is important when we use segments as building blocks or modules to construct complete algorithms: the postconditions of previous modules must satisfy the preconditions of later modules, or the algorithm won't work. (We formalize these concepts later in this chapter by introducing **subalgorithms**.)

If we look at each module as a group of instructions that in many ways is independent of the instructions that may surround it, we obtain an intuition about the nature of a modular problem solution. Let's consider the algorithmic module to list and sum an end-of-file data list:

```
set sum to 0
write, "The data list is:"

while there is data left do
   read, x
   write, x
   set sum to sum + x
end while loop
```

ALGORITHM EXAMPLE 7–3

The precondition for this module is that a data list exist. The essential postcondition is that the value of variable sum is the total of all the data items (if any). To use the module, we do not care that there is another postcondition—that the variable x contains the last data value. We are interested only in sum, because, presumably, it is used later in another module within our complete algorithm. The variable x is required within the summing module to transmit each data value to the assignment instruction set sum to sum + x, and for no other purpose. We call such a variable a **local variable** because it is needed only in a localized part (module) of the complete algorithm.

At this point we can define more exactly what we mean by *module*. A **module** should implement a single conceptual computation (summing, averaging, finding the minimum or maximum, etc.). Usually, a module requires only a few preconditions and produces only a few *essential* postconditions. Nonessential postconditions are values that are not useful in the remainder of the algorithm, such as those in local variables.

We stress the modularity of algorithms (or the structuring of an algorithm by modules) to decrease conceptual complexity. As problems become harder to solve, the algorithmic solutions usually become longer and more difficult to construct. If we insist on writing large, difficult algorithms in the same way as we do small, easy ones, we eventually find a problem whose algorithm seems so complex that we can't handle it—the number of details overwhelms us. If, on the other hand, we structure the algorithm into modules, we worry about only one concept at a time: implementation of the current module. Also, we restrict the number of details we must handle: we need worry about only the essential preconditions we need (hopefully few), the current module's local variables, and the essential postconditions.

Modular algorithm design relies on another algorithm design technique, known as **top-down** or **hierarchical** design, that requires us to break a solution down into modules. At first, the modules implement very high-level concepts—when the problem is complicated enough to have such concepts. Each module is **refined** or restated in terms of submodules; each submodule implements a high-level concept, and is subsequently refined into sub-submodules; and so on, until we reach the conceptual level of the machine: assignment, read, write, if-then-else, etc. This is called **stepwise refinement**.

At each stage in stepwise refinement, of course, the modules must fit together; that is, the postconditions of previous modules must satisfy the preconditions of the current module. The idea behind modular and hierarchical design is like the military concept "divide and conquer." If we divide the "enemy" (the complexity of an algorithm) into small factions (modules), we can directly attack the factions that are small enough to be "conquered" easily (written in terms of assignment, read, write, if-then-else, etc.), or we can further subdivide factions (refine modules) until we can easily solve them. However, the subdividing must be done at carefully chosen points: each module is chosen to implement only one idea or concept. Furthermore, we choose modules so that they are neither too big (too much at once) nor too small (not enough to justify a separate module).

"How can one know, being a novice, how to choose modules properly?" We can use two rules of thumb. A module is about the right size if it can be refined in more than one line, but fewer than 50 lines (or 1 page), whichever is smaller.[1] Also, a module implements a single concept, and therefore has only a few preconditions and essential postconditions—generally fewer than 5.

Before we define a module further, we need another term: *interface*. According to most dictionaries, an interface is the surface between two parts of matter or space, and forms their common boundary. In algorithms and other computer usage, an **interface** is the boundary between, or meeting of, two modules. An interface is achieved by meshing the postconditions from previous modules with the preconditions of the current module. The term *interface* can be used with both software and hardware modules.

7.2. SUBALGORITHMS AND PARAMETERS

Modules may be inserted directly within a larger algorithm or module, or may be written as separate, independent modules (if the latter, they are called **subalgorithms**). By *separate* and *independent*, we mean that no variables are shared by an algorithm and its subalgorithm. In other words, the names of variables in both an algorithm and a subalgorithm are independent. We may use the same name in both, but the name is associated with one memory cell in the algorithm and a different one in the subalgorithm.

To make use of a subalgorithm, we need to specify three things: (1) how to ask to use the subalgorithm from an algorithm, (2) how to supply or obtain the variables or values that specify the pre- and postconditions necessary for the subalgorithm to work, and (3) how to describe the subalgorithm.

[1]One page of 50 lines eliminates the distraction of flipping pages and keeps the module small enough for us to understand.

Since subalgorithms are separate, independent modules, we also need a way to refer to them from within another module; and we can **call** or **invoke** them if we give them names. In the algorithms we write here, we allow ourselves to say such things as:

read and sum the data list . . .
read and find the smallest value in the data list . . .

and so on, where each of these high-level statements is refined by writing a subalgorithm to accomplish the required task.

A subalgorithm *call*, another kind of control structure, diverts the sequential execution to the named subalgorithm. After the subalgorithm ends, execution reverts to the instruction that follows the subalgorithm call, and the instructions read and write are examples of this. The read and write performed by the machine is much more primitive than we have been expecting. When we say read or write in an algorithm, we call on a subalgorithm that arranges to do the reading and writing for us (without our worrying about the details demanded by the machine). Shortly, we will see examples of this.

Since a separate, independent module might need certain values set as preconditions before it will execute properly, and to set certain values as postconditions as it executes, we need a mechanism that allows a subalgorithm to inspect and change the values of variables from the calling module. In a read and sum the data list subalgorithm, the precondition is existence of a data list and the postcondition is a variable in which the subalgorithm sums the data list. Because the list is always available to any module that uses the read instruction, we do not need a special way to handle the input. (Similarly, the write instruction makes the output list available.) The variable in which the calling module wants to find the result (postcondition) of read and sum the data list must be explicitly given to the subalgorithm, since the variables in the caller are not normally accessible to the subalgorithm. For example, we might say:

```
Start Algorithm
   read and sum the data list receiving TOTAL
   write, "The total of the values in the data is ", TOTAL
   stop

End Algorithm
```

ALGORITHM EXAMPLE 7–4

where receiving TOTAL means that the algorithm expects to find a value, received from the subalgorithm, in the variable TOTAL. If we need to give the values of variables to a subalgorithm as a precondition, we would have said given variable1, variable2, etc. Thus we distinguish pre- from postconditions.

Sometimes a variable serves as *both* a pre- and postcondition, and should be listed in both the "givens" and the "receivings." In general, the form of a call to a subalgorithm (as we will use it in this text) is

name of the subalgorithm,
given a list of preconditions and
receiving a list of postconditions

The pre- and postcondition variable names in a call are **actual parameters**. As with any standard form we use, the student is free to formulate a different way of stating it, as long as the meaning is clear and is equivalent in function. A subalgorithm call must stipulate the name of the subalgorithm to be called and the list of pre- and postconditions.

Finally, we need to describe the subalgorithm itself. A subalgorithm starts with a heading that identifies the name, the list of preconditions, and the list of postconditions for the subalgorithm. The heading parallels the features that appear in the subalgorithm call. For example, the subalgorithm to sum a data list might start with the heading

Subalgorithm to read and sum the data list returning SUM.

Thus the heading starts with "Subalgorithm to . . .", followed by its name, and we say (for clarity) "returning" (rather than "receiving") to introduce the list of postconditions. You might think that there has been a mistake: in the call, the postcondition was named TOTAL, but here we called it SUM. This is no mistake; when we said that a subalgorithm is separate and independent, we meant it. The names of variables and parameters in a subalgorithm have nothing to do with the names of variables or parameters in the calling module—or in any other module. The name SUM in the heading is taken to mean the same as TOTAL in the subalgorithm call because *they are in the same position in the list of postconditions.*

The separate naming for subalgorithms and other modules allows us to be unconcerned with the names that other programmers use. Most programming languages provide a subprogram (equivalent to a subalgorithm) to find the square root of the value of a variable. Let's assume that separate naming does *not* hold and that the person who wrote the square root subalgorithm used a variable called X. If the names in the subalgorithm and the calling algorithm have to be the same, then we would have to call the variable X, too. Should we avoid naming a variable or a parameter X because someone else wants to use this name? For that matter, what *other* names might we have to avoid? We begin to see that for the small algorithms we have written so far, separate naming might seem an inconvenience, but in large algorithms it is necessary.

If the name of the pre- or postcondition variable in the call for the subalgorithm is not the same as that of the variable used for the same purpose in the heading of the subalgorithm, how do we know what is meant? The position of names in the pre- and postcondition lists determines a correspondence that connects the names.

For example, if we have an algorithm to compute a statistic called the "*variance*," we might use a variable named Variance. We might also want the standard deviation (which is the square root of the variance) in a variable called StdDeviation. In our algorithm, we would write:

```
Find a square root given Variance and
        receiving StdDeviation
```

The subalgorithm to find the square root might be written with this heading:

```
Subalgorithm to find a square root
      given X and returning SQRT
```

In this case, when the subalgorithm begins to execute, wherever it refers to X it gets the value that is stored in the variable Variance. Similarly, when the subalgorithm stores a value into SQRT, it stores the value into StdDeviation.

The names of parameters in subalgorithm headings (called **formal parameters**) allow us to refer to the values and memory cells given by or returned to the calling algorithm—without knowing what they were named in the calling algorithm. Formal parameter names can be looked at as pseudonyms or alternative names for the real names supplied by the calling algorithm. Parameters in the subalgorithm call are called **actual parameters** because of this.

7.3. PARAMETER PASSING TECHNIQUES

Programming languages use various techniques to accomplish *parameter passing*, but two techniques are common and we will use both of them.

The first technique called **pass by value**, is used only to "pass" values to parameters that are preconditions. In essence, the formal parameter name is associated with a memory cell, just like a variable. When the subalgorithm is called, a copy is made of the value of the actual parameter and put into the corresponding formal parameter. The subalgorithm can then use the name of the formal parameter, and even change the value it contains, but *no change can be made to the value of the actual parameter* because the subalgorithm has only its private copy of the parameter, not the parameter itself. When the subalgorithm ends and we resume executing the calling algorithm, the value of the actual parameter cannot have changed.

The second technique, called **pass by reference**, is used with parameters that are postconditions, or both pre- and postconditions. In essence, the formal parameter name becomes, temporarily, the name we use to refer to the same memory cell associated with the actual parameter. When we change the value referenced by the formal parameter name, we change the value in the memory cell that's referenced by the actual parameter. When we finish the subalgorithm and return to the calling algorithm, the values of the actual parameters that were passed as postconditions remain set to the values assigned to them in the subalgorithm. This is exactly what "pass by value" will not do.

A simple example will clarify these ideas. Suppose we have an artificial algorithm that calls an artificial subalgorithm ("artificial" in the sense that they are constructed to show how something works but do nothing else worthwhile):

```
Start Algorithm

  set x to 5
  set y to 10
  change the parameters given x and y and receiving y
  write, x, y
  stop

End Algorithm

Subalgorithm to change the parameters given a and b
                                            and returning b.

  set a to b
  set b to 0
  write, a, b

End Subalgorithm
```

ALGORITHM EXAMPLE 7–5

What is printed by the algorithm/subalgorithm? To trace the execution, we set the values of x and y to 5 and 10 respectively. When we call the subalgorithm, we must see that x in the call corresponds to a in the subalgorithm heading, that they are preconditions only, and so are "passed by value." Similarly, we must see that y and b correspond, are both pre- and postconditions, and so are "passed by reference." A copy of x is made into a (so a also has the value 5), and the name b is made to refer to the same memory cell as y. (If we should look, b has a value of 10).

Now we begin the subalgorithm body. Set a to b causes a to be changed from 5 to 10 (b has the value 10). This causes no change to x, since the parameter is "passed by value," and a is *not* the same memory cell as x, but only started out with the same value.

The next step, set b to 0, also causes y to be set to zero, since b is another name for the same memory cell used in the calling algorithm under the name y (y is "passed by reference").

If we were to use the names x or y in the subalgorithm or the names a or b in the calling algorithm, these names would refer to local variables and *not* to the names in the other module.

7.4. AN EXAMPLE OF MODULAR PROGRAMMING

We are considering, let's say, buying a personal computer (PC), and we know that they have various options. We need the basic computer unit, a CRT display, a printer, perhaps some add-on memory, and software packages. Because there are many brands, we go to local computer stores and survey the market, and return with a list of prices for each item for each PC. Then we write an algorithm to add up the prices for each PC system and print out the name of each component, its price, and the system totals. Our data looks like this:

```
Computer Shack Ziggy-V
Basic-Unit          670.0
CRT                 80.0
16K-memory          125.0
OMNICALC            175.0
Printer             215.0
End                 0.0
Itty Bitty Machines Pico-I
Basic-Unit          500.0
CRT                 95.0
64K-memory          350.0
QUASICALC           25.0
Printer             189.95
End                 0.0
Sushiyama All-In-One
Basic-Unit          995.0
CRT-included        0.0
Printer-included    0.0
SUSHICALC           0.0
End                 0.0
```

Each of these three groups of data starts with a line that specifies the manufacturer's name and model. Then a series of lines details the "extras" we need: the names of the extras and their prices. Each group ends with the sentinel line that has the name "End" and a price of zero.

As we put the data for several groups together, one after another, to form the data for our algorithm, we notice that the data has two levels of *structure*. In the first level, the breakdown is by computer manufacturer/model; that is, the coarse structure of the data list is a sequence of groups of data about specific computers. Within the first (coarse) level of structure is a second (finer) level in which the breakdown is by computer component.

The structure of our algorithm should parallel the structure of the data: the first level of our algorithm should treat the data as a list of computers. We ignore, for the moment, the fact that the data for each computer is a sentinel value data list that details the components of each computer. With this in mind, we see that the list of computers is given as an end-of-file data list:

```
Computer Shack Ziggy-V and details
Itty Bitty Machines Pico-I and details
Sushiyama All-In-One and details
```

Therefore, our algorithm should be:

```
Start Algorithm

  while there is data left do
    process one computer
  end while

End Algorithm
```

ALGORITHM EXAMPLE 7–6

Having provided for the possibility that there may be more than one computer in the data list, we have reduced the problem to process one computer: the first level of refinement in a hierarchical design. Process one computer must become a subalgorithm we need to state as a further refinement. The data list for each computer is a sentinel value list. We need to sum up the prices; so we recall the standard algorithm to sum a sentinel value list and make it into a subalgorithm:

```
Subalgorithm to process one computer

  set sum to 0
  read, value
  while value is not the sentinel do the following
```

```
        set sum to sum + value
        read, value
    end while
    write, "The sum is: ", sum

End Subalgorithm
```

ALGORITHM EXAMPLE 7–7

With this as a starting point, let's see what modifications we must make so that this "standard" subalgorithm conforms to our requirements. The first thing subalgorithm 7–7 does is to read a single data value, and the first thing on the data list for a computer is a title line. After that, each line contains two items: a component and a price. Let's fix these two problems:

```
Subalgorithm to process one computer

    set sum to 0
    read, title
    read, component, price
    while component ≠ 'End' do the following
        set sum to sum + price
        read, component, price
    end while
    write, "The sum is: ", sum

End Subalgorithm
```

ALGORITHM EXAMPLE 7–8

In addition to fixing the read statements to reflect the arrangement of the data, we fixed the check for a sentinel value and changed set sum to sum + price. But we are not yet finished.

The subalgorithm writes only one line: The sum is: _______ (the blank is filled with the total system cost). We are supposed to write out the title line and the names and prices of each component. We should also change the output line from The sum is: _______ to a more appropriate message, like Total system cost is: _______ .

```
Subalgorithm to process one computer

    set sum to 0
    write a blank line
    read, title
```

```
write, title
write a blank line
read, component, price
while component ≠ 'End' do the following
   write, "       ", component, price
   set sum to sum + price
   read, component, price
end while
write a blank line
write, " Total system cost is: ", sum

End Subalgorithm
```

ALGORITHM EXAMPLE 7–9

In this version, we write out the title immediately after we read it in. This is appropriate since we need to write out the title, and doing it immediately after we read it works and minimizes the number of variables we have to worry about later (once it is written out, we don't need the title anymore). We also write out blank lines and add extra spaces inside the quoted messages to make the output look nicer. We do not need parameters with this subalgorithm. The only precondition is the input list and the only postcondition is the output list.

If we translate the main algorithm and subalgorithm into a programming language and run the program on the data below, we should get similar output. The input list is

```
Computer Shack Ziggy-V
Basic-Unit          670.0
CRT                 80.0
16K-memory          125.0
OMNICALC            175.0
Printer             215.0
End                 0.0

Itty Bitty Machines Pico-I
Basic-Unit          500.0
CRT                 95.0
64K-memory          350.0
QUASICALC           25.0
Printer             189.95
End                 0.0
```

```
Sushiyama All-In-One
Basic-Unit          995.0
CRT-included        0.0
Printer-included    0.0
SUSHICALC           0.0
End                 0.0
```

The output generated by the algorithm/subalgorithm is

```
Computer Shack Ziggy-V

    Basic-Unit          670.0
    CRT                 80.0
    16K-memory          125.0
    OMNICALC            175.0
    Printer             215.0

    Total system cost:  1265.0

Itty Bitty Machines Pico-I

    Basic-Unit          500.0
    CRT                 95.0
    64K-memory          350.0
    QUASICALC           25.0
    Printer             189.95

    Total system cost:  1159.95

Sushiyama All-In-One

    Basic-Unit          995.0
    CRT-included        0.0
    Printer-included    0.0
    SUSHICALC           0.0

    Total system cost:  995.0
```

Since we now have an algorithm to total the prices for each system, we could also have it determine the lowest cost system. Currently, our main algorithm is

```
Start Algorithm

  while there is data left do
    process one computer
  end while

End Algorithm
```

ALGORITHM EXAMPLE 7–10

We also have an algorithm to find the smallest value in a data list (see section 3.2):

```
Start Algorithm

  set smallest to a ridiculously large number
  while there is data left do
    read, value
    if value < smallest
      then
        set smallest to value
    end if
  end while
  write, "The smallest value is: ", smallest

End Algorithm
```

ALGORITHM EXAMPLE 7–11

Now we must *merge* these two algorithms so that we do two things: find the total system cost for each PC and find the smallest total system cost. We can't simply put the two algorithms end to end and execute them sequentially, because the first one would read all the data. The second one would do nothing, since no data would be left for it to work with.

We must use the features that correspond to one another in these two algorithms to interleave the two algorithms. We must not duplicate the features they have in common, and we must include the features they do *not* have in common. It is important to note that both algorithms expect to read data from an end-of-file data list. If they expected different types of data lists, we would have to decide on one type and then make sure that both conform to the standard structure for reading that type.

The standard structure for reading an end-of-file data list is

```
Start Algorithm

  [initialization]
  while there is data left do
    read, value
    [process value]
  end while
  [summary processing]

End Algorithm
```

ALGORITHM EXAMPLE 7–12

Start Algorithm
  [initialization]
  while there is data left do
    read, value
    [process value]
  end while
  [summary processing]
End Algorithm
	Start Algorithm
	  [not needed]
	  while there is data left do
	    process one computer
	
	  end while
	  [not needed]
	End Algorithm

FIGURE 7–1 Comparing two algorithms: Read with Process

Start Algorithm
  [initialization]
  while there is data left do
    read, value
 	
    [process value]
 	
  end while
  [summary processing]
End Algorithm
	Start Algorithm
	  set smallest to a ridiculously large number
	  while there is data left do
	    read, value
	    if value < smallest
	      then
	        set smallest to value
	    end if
	  end while
	  write, "The smallest value is: ", smallest
	End Algorithm

FIGURE 7–2 Comparing two algorithms: Read with Smallest

In Figure 7–1, we see that the algorithm to find the total system costs of a list of computers has nothing corresponding to [initialization]. The algorithm to find the smallest has only one statement that corresponds to [initialization]:

set smallest to a ridiculously large value

Our combined algorithm should then have an [initialization] section that contains only the statement to set smallest to a large value.

The while loop appears in both algorithms, as well as in the standard module to read end-of-file data lists. In our combined algorithm, the while loop would appear only once:

```
Start Algorithm

   set smallest to a ridiculously large number
   while there is data left do
      read, value
      [process value]
   end while
   [summary processing]

End Algorithm
```

ALGORITHM EXAMPLE 7–13

Next, we consider what corresponds to read, value. In the "smallest" algorithm, we have read, value. In the "system cost" algorithm, however, we have process one computer. At this point we might be fooled into thinking that process one computer should correspond to [process value], which is the next item on our checklist (standard algorithm to read end-of-file data). Process one computer does not correspond to [process value].

The subalgorithm process one computer reads all the data associated with one computer, as well as adding the prices together and writing things out. In that case, how do process one computer and read, value correspond? To understand this, we must step back for a more general overview. In the combined algorithm, what should value stand for? Since we are looking for the smallest total system cost, value should stand for the total system cost of the last computer processed by process one computer. This means that we don't read in our combined algorithm but, as in the "system cost" algorithm, call on process one computer to obtain the total system cost for use in the remainder of the combined algorithm. We need to modify the subalgorithm to return the total system cost as a postcondition. We must also modify the call of process one computer to receive the total system cost. We now have

```
Start Algorithm

  set smallest to a ridiculously large number
  while there is data left do
    process one computer, receiving totalcost
    [process value]
  end while
  [summary processing]

End Algorithm
```

ALGORITHM EXAMPLE 7–14

The next item to be checked is [process value]. The "total cost" algorithm has nothing to correspond to [process value]. The "smallest" algorithm has the if statement; so, after changing value to totalcost to agree with the postcondition received from process one computer, we have

```
Start Algorithm

  set smallest to a ridiculously large number
  while there is data left do
    process one computer, receiving totalcost
    if totalcost < smallest
      then
        set smallest to totalcost
    end if
  end while
  [summary processing]

End Algorithm
```

ALGORITHM EXAMPLE 7–15

Similarly, only the "smallest" algorithm has anything to correspond to the [summary processing] section of the standard end-of-file module. After changing the message in the write statement to be more meaningful in the current context, we have

```
Start Algorithm

  set smallest to a ridiculously large number
  while there is data left do
```

```
    process one computer, receiving totalcost
    if totalcost < smallest
      then
        set smallest to totalcost
    end if
  end while
  write, "The smallest total system cost is: ", smallest

End Algorithm
```

ALGORITHM EXAMPLE 7–16

The output—after all our work—has changed by the single line:

```
The smallest total system cost is: 995.0
```

We now have to find the name of the system whose total cost is 995.0. Perhaps we should also return the title line from the process one computer subalgorithm so we can write that out, too—only the title line for the system with the smallest total cost, however. In fact, wherever we use the variable smallest, we must include statements to handle a new variable: SmallestTitle. We should initialize SmallestTitle to a message that indicates there is no data (if there isn't), then update it where we update smallest. Our final algorithm and subalgorithm are

```
Start Algorithm

  set smallest to a ridiculously large number
  set SmallestTitle to "There was no data to be read."

  while there is data left do
    process one computer, receiving totalcost and title
    if totalcost < smallest
      then
        set smallest to totalcost
        set SmallestTitle to title
    end if
  end while

  write, "The smallest total system cost is: ", smallest
  write, "The system with the smallest total cost is: ",
                    SmallestTitle

End Algorithm
```

ALGORITHM EXAMPLE 7–17

```
Subalgorithm to process one computer,
            returning sum and title

  set sum to 0
  write a blank line
  read, title
  write, title
  write a blank line
  read, component, price
  while component ≠ 'End' do the following
    write, "      ", component, price
    set sum to sum + price
    read, component, price
  end while
  write a blank line
  write, " Total system cost is: ", sum

End Subalgorithm
```

ALGORITHM EXAMPLE 7–18

7.5. SUMMARY

In this chapter we have discussed the *modularity* of algorithms, the concepts of *preconditions* and *postconditions*, the writing and use of *subalgorithms*, and the use of *parameters*.

A module was described as implementation of one idea, such as computing a sum. We showed how the values of variables are used and changed in the course of executing a module, and we defined the notions of pre- and postconditions for a module. We also described the *step-wise refinement* technique, in which we specify successively more detailed instructions. This is also referred to as *hierarchical design*.

The isolation of a module and the explicit statement of its pre- and postconditions led us to the notion of a *subalgorithm*. We defined the subalgorithm call mechanism and described two ways of supplying *parameters*: *pass by value* and *pass by reference*. The parameters of a subalgorithm are the explicit statement of the subalgorithm's pre- and postconditions.

We ended the chapter by solving a simple problem: Find the cheapest personal computer, given a list of computers detailed by component. In doing this, we *merged* two algorithms by identifying the similarities in their structure.

7.6. EXERCISES

7.01. Ms. Mia Usticar, who is interested in buying a used car, has narrowed her choice to several cars that seem to be in excellent condition. The deciding factor is the gas mileage; she will buy the car with the highest MPG.

The dealer has given Ms. Usticar the data for calculating the average MPG of each car, consisting of the mileage driven on various amounts of gas on each of four occasions over the last several months. The data file contains a single-letter code for each car, followed by the four data pairs, each on a separate line. Each data pair has the mileage (to the nearest mile), followed by the gallons of gas used (to the nearest tenth of a gallon). The end of the data file is indicated by a car code, '#'.

Compute and print the average MPG for each car, as well as the car that Ms. Usticar will select (and its MPG).

Sample Input

```
V
102  3.1
271  9.3
189  5.9
343 13.1
Z
288 11.6
132  4.8
 91  4.1
175  7.0
#
```

Sample Ouput

```
Car V      Miles      Gallons
            102          3.1
            271          9.3
            189          5.9
            343         13.1

Miles per Gallon = 28.82

Car Z      Miles      Gallons
            288         11.6
            132          4.8
             91          4.1
            175          7.0
```

Miles per Gallon = 24.95

Car V had the highest MPG, 28.82

7.02. You have just been hired as a programmer for ACME Shipping to write a program to produce bills of lading for shipments (a bill of lading is a shipping list, with shipping costs detailed).

There are data for several bills of lading, and several lines of data for each bill of lading. Each line of data has a shipping zone, part number, and weight. The last line on each bill of lading has a shipping zone, 'S', and there are four shipping zones, with charges as follows.

Zone	*Base Rate (per item)*	*Poundage (per pound)*
A	$ 5.00	$ 0.05
B	$ 7.50	$ 0.075
C	$ 10.00	$ 0.075
D	$ 10.00	$ 0.10

Sample Input

```
A 12345  55.32
C 98765  27.00
D 83475 105.00
S     0  0.00
S     0  0.00
B 39483  78.50
A 23948 139.00
S     0  0.00
```

Sample Output

ACME Shipping Company, Inc.
1234 W. 5th St.
Lawrence, Ks 66044

Item	Part #	Weight	Shipping Zone	Shipping Cost
1	12345	55.32	A	7.77
2	98765	27.00	C	12.03
3	83475	105.00	D	20.50

Total for the 3 items in this bill: $ 40.29

ACME Shipping Company, Inc.
1234 W. 5th St.
Lawrence, Ks 66044

Item	Part #	Weight	Shipping Zone	Shipping Cost

***** WARNING: There is an empty bill.

ACME Shipping Company, Inc.
1234 W. 5th St.
Lawrence, Ks 66044

Item	Part #	Weight	Shipping Zone	Shipping Cost
1	39483	78.50	B	13.39
2	23948	139.00	A	11.95

Total for the 2 items in this bill: $ 25.34

7.03. Write a program to grade simple arithmetic exams.

1. Read exam questions from the data file, whose first line contains the number of questions in each test. The remaining lines contain the arithmetic problems, in the form:
   operand1 operator operand2 = answer
   The input list ends at the end-of-file.
2. The valid operations are ' + ', ' − ' and 'X' (addition, subtraction, multiplication). Determine if an answer is correct or incorrect. Any question with an invalid operation is counted incorrect, and an appropriate warning message is written.
3. Print out each test as shown below. You may assume
   1. Exactly 1 blank on either side of the operation and the equal sign, and the operands (except the sentinel) are between 0 and 99.
   2. The data file will not stop in the middle of a test.

Sample Input

```
2
1 + 1 = 2
3 × 5 = 15
2 + 2 = 4
4 × 3 = 11
```

Sample Output

```
Test # 1
  1 + 1 = 2 is correct.
  3 × 5 = 15 is correct.
Number of correct answers = 2
Number of incorrect answers = 0
```

Test # 2
  2 + 2 = 4 is correct.
  4 × 3 = 11 is incorrect.
Number of correct answers = 1
Number of incorrect answers = 1

7.04. Write a program to determine the winner of a race. There are several runners in every race; in addition, there are two timekeepers for each runner. One timekeeper records the time in which a runner runs the entire race; the other timekeeper records the time for each lap (there may be a different number of laps in each race, but all the runners in every race run the same number of laps).

The first line of the data file contains the number of laps in a race. The data for each runner starts with the line that has the initials of the runner (up to 3 letters). The next line has the time recorded by the first timekeeper, followed (on successive lines) by the times recorded by the second timekeeper. The input list ends at the end-of-file.

Compare the time recorded by the first timekeeper to the sum of lap times given by the second timekeeper. In addition, determine the winner of the race. (To avoid confusion, the winner is determined by the first timekeeper's time.)

Sample Input

```
3
QIK
120.3
 35.1
 40.2
 45.0
FAS
118.2
 30.1
 45.2
 42.5
SLO
253.1
 98.6
100.2
 54.3
```

Sample Output

```
This race has 3 laps.

Runner: QIK total race time: 120.3

    Lap times:
              35.1
              40.2
              45.0
              ___
    total    120.3

Both timekeepers agree.

Runner: FAS total race time: 118.2

    Lap times:
              30.1
              45.2
              42.5
              ___
    total    117.8

The timekeepers disagree by 0.4 minutes.

Runner: SLO total race time: 253.1

    Lap times:
              98.6
             100.2
              54.3
              ___
    total    253.1

Both timekeepers agree.

And the winner of the race is . . .

        FAS
with a time of 118.2 minutes
```

7.05. Write a program that will read a list of numbers and produce the prime factors for each input number. The input list will be terminated with a value of 1 or less.

(A prime number is divisible only by itself and 1. You can tell if a number is a factor of another if the factor divides the number evenly; that is, if the remainder is 0.)

One way to find all the prime factors of any number is to start at 2 and work up, testing whether each number in between is a factor. To make sure that this yields only the prime factors, divide the original number by each factor as many times as possible before you continue.

Sample Input

```
 5
91
72
 1
```

Sample Output

```
Prime factors of  5: 5
Prime factors of 91: 7 13
Prime factors of 72: 2  3
```

Possible extension: For each number, also print out its greatest (not necessarily prime) factor (not including the number itself).

Sample Output

```
Greatest factor of  5 is  1
Greatest factor of 91 is 13
Greatest factor of 72 is 36
```

8
Composite Data Types

8.1. PROBLEMS THAT CAN'T BE DONE WITH SIMPLE TYPES

The problems we have solved thus far have a very particular property: there was no solution in which we had to remember a previous data value after we processed it. For example, in the algorithm to find the largest value in the data list, we read in a value, checked it against the current "champion," and either kept the old champion and ignored the new value or kept the new value as the champion and discarded the old. Therefore we needed only two variables: one to store the current data value and one to store the current champion. We had no need to remember the first or second (or any other previous) data value in order to see if the current value should replace the champion.

Let us pose a simple problem and write an algorithm for it, using only the techniques we know: Read in a list of values and print them out in reverse order. Given the input list 10, 20, 30, 40, 50, we should write out 50, 40, 30, 20, 10. (This sounds easy.) We could write the algorithm

```
Start Algorithm

   read, A
   read, B
   read, C
   read, D
   read, E
```

```
    write, E
    write, D
    write, C
    write, B
    write, A
    stop

End Algorithm
```

ALGORITHM EXAMPLE 8–1

But this has a major drawback: it works only if there are exactly five values in the data list. For the algorithm to work for exactly five values, we need at least five variables. To write out the last value (which we read in first) on the output list, we need to store all the values that precede it in separate memory cells so we can write them out later. Our previous method of recycling a single variable will not work in solving a problem like this. Instead, we need a variable, or at least a memory cell, for each data value.

The objection to Algorithm 8–1—that it will not work for two or three or four values, but only for five–can be removed if we can fix the algorithm to work for data lists of length zero through five. Remember the number of values, then write out only those variables into which we read values.

```
Start Algorithm

    set count to 0
    if there is data
      then
        set count to 1
        read, A
    end if
    if there is data
      then
        set count to 2
        read, B
    end if
    if there is data
      then
        set count to 3
        read, C
    end if
    if there is data
      then
```

```
        set count to 4
        read, D
  end if
  if there is data
     then
        set count to 5
        read, E
  end if
  if count ≥ 5
     then
        write, E
  end if
  if count ≥ 4
     then
        write, D
  end if
  if count ≥ 3
     then
        write, C
  end if
  if count ≥ 2
     then
        write, B
  end if
  if count ≥ 1
     then
        write, A
  end if
  if count = 0
     then
        write, "The data list is empty."
  end if
  stop

End Algorithm
```

ALGORITHM EXAMPLE 8–2

If you trace this algorithm with a data list of 8, 9, 10, you obtain an output of 10, 9, 8. The algorithm works because before we read in data, we check to see if any data is left to be read. If there is, we read a single value into a new variable, then set count to the number of values we have read so far. If no data is left, we skip the remaining read and assignment statements because the condition in the remaining if there is data checks is false. If there were three values in the data list, we would read the values into variables A, B, and C and count would be equal to 3. Since each write statement is protected by an if statement that checks the value of count, we will skip over the write, E and write, D because count is only 3 and is not greater than or equal to 5 or 4, respectively. We then write out the values in the variables C, B, and A, in that order, because the value of count is greater than or equal to 3, 2, and 1, respectively.

Algorithm 8–2 is 50 lines long and handles only data lists of length—at most, five. If we give it a data list of length six or longer, it will process only the first five values and ignore the rest, which is poor style. The algorithm should detect such an error and write a message. Data lists that are several hundred (even several thousand values long) are not uncommon. Using the technique above, and a data list that is at most 200 values long, we would need 200 variable names, 200 if there is data statements, and 200 if count ≥ _______ statements. In other words, our algorithm is 1805 lines long and is growing at the rate of nine lines each time we increase by one the length of the longest data list the algorithm will correctly handle.

Because this problem has elbowed its way into our thinking, let's see what such a long algorithm would look like. We need 200 variable names, and for convenience we'll call them X1, X2, . . . , X200.

```
Start Algorithm

  set count to 0
  if there is data
    then
      set count to 1
      read, X1
  end if
  if there is data
    then
      set count to 2
      read, X2
  end if
  if there is data
    then
      set count to 3
      read, X3
  end if

              . . .

  if there is data
    then
```

```
        set count to 199
        read, X199
end if
if there is data
    then
        set count to 200
        read, X200
end if

if count ≥ 200
    then
        write, X200
end if
if count ≥ 199
    then
        write, X199
end if

                  . . .

if count ≥ 2
    then
        write, X2
end if
if count ≥ 1
    then
        write, X1
end if
if count = 0
    then
        write, "The data list is empty."
end if
stop

End Algorithm
```

ALGORITHM EXAMPLE 8–3

The fact that, to handle the complexity of naming 200 variables, we numbered them consecutively, will be important in the following section. It is also important (in keeping the algorithm understandable) that the number of each variable is the same as the count of the items read in to that point. Our naming conventions give us a rational way to remember which variables we used and the order in which we used them.

8.2. WAYS OF AGGREGATING DATA

As we saw in the last section, problems are difficult (or at least tedious) when we are allowed to use variables only in the sense we have defined them: a variable name associated with a *single* memory cell. We must now consider a relaxation of that definition. Virtually all programming languages allow at least one type of variable name that can be associated with several memory cells, and the most common aggregation is called an **array**, which consists of a name, a sequence of index values, and a collection of memory cells. This name allows us to refer to the collection of memory cells as a *unit*. This sequence contains one distinct value for each memory cell, and is always contiguous and countable. These index values allow us to select one element of an array from the collection by giving both the name of the array and one of the index values.

This process is like addressing a letter to a Post Office box: the collection of boxes is analogous to an array and the box number is analogous to an index value. Examples of indexing sequences are 1, 2, 3, 4, 5, or 'a', 'b', 'c', 'd', 'e', or −2, −1, 0, 1, 2, etc. Each of these sequences is contiguous (i.e., there are no gaps in the sequence) and they are countable types: integer and character. (Real values cannot be used in an indexing sequence because it is impossible to have a contiguous sequence of real values: there are always gaps, no matter which real values you pick. However, some programming languages, notably FORTRAN and BASIC, restrict the indexing sequence to integers, and sometimes to sequences of integers starting at 1.)

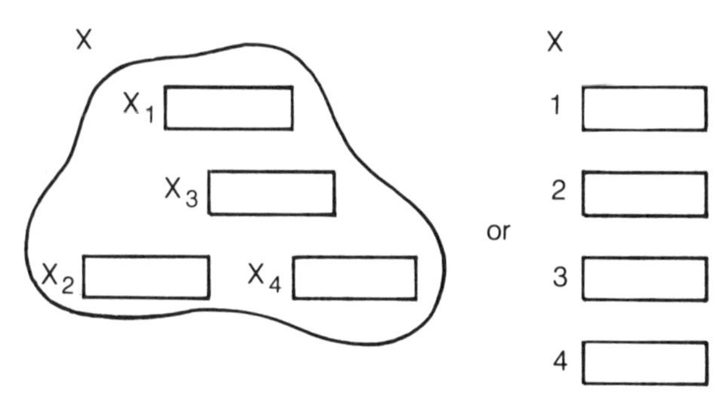

FIGURE 8–1 Array, named X, with 4 Elements

Thus we've seen a way to aggregate homogeneous data. An array is a collection of values, all of which have the same element type (an array is homogeneous): all integers or all reals or all characters, etc. There are circumstances, however, in which we would like to group *non*-homogeneous data—for instance, if we are asked to write an algorithm to keep track of books in a library. To represent a book, we need at least these pieces of information:

Title	(string of characters)
Author	(string of characters)
Publisher	(string of characters)
DateOfPublication	(string of characters)
NumberOfPages	(integer)
Hardback	(true or false)
ISBN	(string of characters)
LibraryOfCongressRef	(string of characters)
PurchasePrice	(real)

FIGURE 8–2 Description of a Book Record

Each of these nine pieces of information is an attribute of a book, reducible to the forms of character string, integer, real, and boolean values. Because they are not all of the same data type, we can't put them into an array; so we use a new aggregation, a **record**, to store these values. This is a new data type, a *book record*. Each name of an item of information (e.g., title or author) is called a **field name**.

If we declare a variable, say TEXT to be of the type book record, it might look like Figure 8–3.

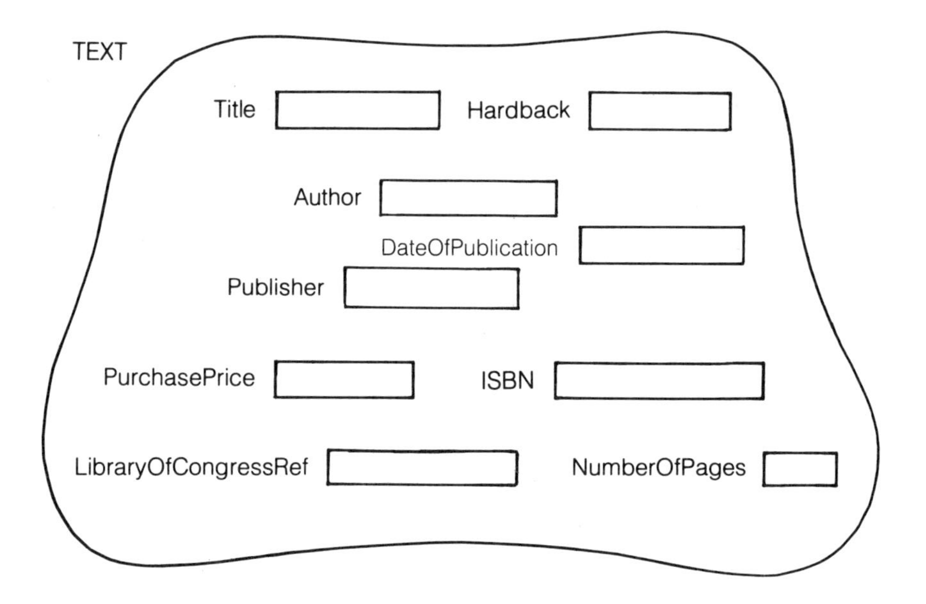

FIGURE 8–3 Variable of Type Book Record

8.3. ARRAYS AND THEIR USES

Each of the sequences 1, 2, 3, 4, 5; 'a', 'b', 'c', 'd', 'e'; and −1, −2, 0, 1, 2 contains five elements, and we can make an array with five memory cells with such sequences. Each member of the indexing sequence is associated with one, and only one, memory cell. Likewise, each memory cell is associated with one member, and only one, of the indexing sequence.

The benefit from using an array is our ability to refer to a specific memory cell, called an **array element**, by using a variable to contain an indexing value. We refer to a specific array element by using a subscript. For example, if we have an array named VALUES and an indexing sequence 1, 2, 3, 4, the array contains four memory cells that we can refer to by writing: $VALUES_1$, $VALUES_2$, $VALUES_3$, $VALUES_4$. In general, we can set a simple variable *of the same type as the elements of the indexing sequence* to an indexing value, then use the variable as a subscript:

```
set i to 3
set VALUESᵢ to 10.5
```

ALGORITHM EXAMPLE 8–4

This sets the memory cell associated with index value 3 to the data value 10.5; but now we encounter a confusing fact: each memory cell in the array may contain a value, and each memory cell is associated with a member of the indexing sequence. We must keep these two points separate and understand the differences between them. Essentially, two "types" are associated with an array, and one type is that of the data values stored in the memory cells. All elements of the same array contain values from the same data type (integer, real, character, boolean). The other type is that of the indexing sequence, which is always a countable type (integer, character, boolean, but not real). Thus it makes sense to talk of an **index type** and an **element type** when we describe an array.

It is common to use an index type that consists of integers and starts at 1. If we need 200 elements in the array, we choose the indexing sequence 1, 2, 3, . . . , 199, 200. Some programming languages force us to use this sort of indexing sequence; others give us a choice.

We will choose an element type that's based on the type of data values we need to store—just as we chose data types for the simple variables before we "discovered" arrays. A short example will clarify this.

Suppose we have an array named X with an indexing sequence of 1, 2, 3, . . . , 100. We can write an algorithm segment to store zeros into all of these 100 array elements by using a for loop, containing a single assignment statement. We use a for loop rather than a while loop because we know, before we start the loop, that we need to go through the loop exactly 100 times, with the index variable starting at 1 and going to 100.

```
for i ← 1 to 100 do
   set Xi to 0.0
end for
```

ALGORITHM EXAMPLE 8–5

Remember, this segment is equivalent to

```
set i to 0
while i < 100 do
   increment i by 1
   set Xi to 0.0
end while
```

ALGORITHM EXAMPLE 8–6

The for loop sets the index variable i to successive values from 1 to 100. For each setting i, the loop body is executed, so that successive memory cells, from X_1 to X_{100}, are set to 0.0. In this example, the element type of the array X is real.

Let's look at the algorithm by which we reversed a list of values. Now, perhaps, we can write a more reasonable version.

```
Start Algorithm

   set count to 0
   if there is data
      then
         set count to 1
         read, X1
   end if
   if there is data
      then
         set count to 2
         read, X2
   end if
   if there is data
      then
         set count to 3
         read, X3
   end if

               . . .

   if there is data
```

```
    then
      set count to 199
      read, X199
end if
if there is data
    then
      set count to 200
      read, X200
end if

if count ≥ 200
    then
      write, X200
end if
if count ≥ 199
    then
      write, X199
end if

              . . .

if count ≥ 2
    then
      write, X2
end if
if count ≥ 1
    then
      write, X1
end if
if count = 0
    then
      write, "The data list is empty."
end if
stop

End Algorithm
```

ALGORITHM EXAMPLE 8–7

Looking at Algorithm 8–7, we see two patterns: a section of repeated instructions of the form

```
if there is data
  then
```

```
    set count to i
    read, X_i
  end if
```

ALGORITHM EXAMPLE 8–8

where i takes on the values 1, 2, 3, . . . , 200, and another section of repeated instructions of the form

```
if count ≥ i
  then
    write, X_i
  end if
```

ALGORITHM EXAMPLE 8–9

where i takes on the values 200, 199, 198, . . . , 2, 1. We have learned that such patterns suggest the possibility of using loops.

If we use an array, two loops could describe the algorithm to reverse a data list:

```
Start Algorithm

  set count to 0
  while there is data do
    increment count by 1
    set i to count
    read, X_i
  end while

  write, "The reverse of the input list is:"
  write a blank line
  for i ← starting at count and decrementing to 1 do
    write, X_i
  end for

  if count = 0
    then
      write, "The data list is empty."
  end if

  stop

End Algorithm
```

ALGORITHM EXAMPLE 8–10

The while loop continues only as long as there is data. At each pass through the loop, we increment count, set i, and use a new memory cell, X_i. The for loop starts with count as it was when the while loop finished, and executes one pass through the loop for successively decreasing values of i, writing out X_i each time. We have eliminated repeated *writing* sections of algorithm by specifying (in a loop) that the section is to be repeatedly *executed*. We can do this because the variable index, i, when used with the array X, allowed selection of a different array element each time through the loop.

There are two more points about arrays. The first is that an array's size—that is, the length of the indexing sequence—is usually fixed when the computer program that corresponds to an algorithm is written. The actual value of the highest index is not important. What *is* important is that we don't allow the value of a subscript to exceed the highest index value for any given array. This means we must build checks into the algorithm (as we shall see).

The second point is a consequence of the first. Because the indexing sequence is fixed, the number of memory cells allocated to the array is fixed. When we use arrays in programming, we usually choose an indexing sequence that is larger than the largest anticipated data list. This means that some memory cells in the array are used and some are not. We must therefore have a means of determining which have stored values and which do not. Since we usually begin by putting the first data item into the memory cell associated with the lowest index value and proceed to successively higher index values, we need only remember how far we got. We use a variable to remember this index value and call it a **cursor** or **placekeeper**. Index values that are less than or equal to the cursor variable have data in their associated memory cells; index values that are greater than the cursor variable do not. This assumes the convention that we put data into the array in successive index values and do not scatter the data randomly.

8.4. STANDARD SUBALGORITHMS THAT USE ARRAYS

Most of the algorithms that we have written without arrays can be usefully restated as subalgorithms that use arrays. Thus we present three subalgorithms to read the three kinds of data list into an array, subalgorithms to sum and average the values *already stored* in an array, and a subalgorithm to find the smallest/largest of the values *already stored* in an array. Throughout, we will use the indexing sequence 1, 2, 3, . . . , MAXINDEX, where MAXINDEX is an integer large enough to handle a typical data list.

8.4.1. Reading Data into an Array

We saw a way to read data into successive elements in an array in the example to reverse a data list: set a counter variable to zero and increment it as we read in the data values. We then put the first data value from the list into array element 1, the second data value into array element 2, and so on. Here, the counter variable doubles as the cursor or placekeeper and as the subscript. There's no real reason to have the variable i in the reverse-the-data-list algorithm. Any variable with the right properties will do, and it needn't be named i. We start with a standard end-of-file read loop:

```
Start Algorithm

  [initialization]
  while there is data left do
    read, value
    [process value]
  end while
  [summary processing]

End Algorithm
```

ALGORITHM EXAMPLE 8–11

To fit the situation, we modify Example 8–11 by making it a subalgorithm whose only purpose is to read the data into an array (named X in the subalgorithm) and to return the filled array and its last-used index value (the cursor) as postconditions. Therefore, there is no [summary processing].

```
Subalgorithm to read an eof data list filling in an array
          returning array X and its cursor COUNT.

  [initialization]
  while there is data left do
    read, value
    [process value]
  end while

End Subalgorithm
```

ALGORITHM EXAMPLE 8–12

The [initialization] we need is to set the counter to zero, meaning we have read nothing yet. The [process value] section must count the value just read and store it into the array at index position COUNT.

```
Subalgorithm to read an eof data list filling in an array
        returning array X and its cursor COUNT.

  set COUNT to 0
  while there is data left do
    read, value
    increment COUNT by 1
    set X_COUNT to value
  end while

End Subalgorithm
```

ALGORITHM EXAMPLE 8–13

If we trace this subalgorithm, we find that it works fine, up to a point. Because the array has a fixed number of memory cells, it thereby has a fixed highest index value (which we've agreed to call MAXINDEX). For example, if MAXINDEX is 5, but ten data values are on the data list, our subalgorithm will continue past COUNT = 5 and attempt to store the sixth data value into X_6. But X_6 does not exist; so this is a serious error. We must stop the loop before we go beyond the upper bound of the index values. Furthermore, if the subalgorithm fails to read in all the data because of this, we should have the subalgorithm write an error message, informing the user of the error.

After we make these changes, our subalgorithm is

```
Subalgorithm to read an eof data list filling in an array
        returning array X and its cursor COUNT.

  set COUNT to 0
  while (COUNT < MAXINDEX) and (there is data left) do
    read, value
    increment COUNT by 1
    set X_COUNT to value
  end while

  if there is data left
    then
      write, "The data list is too long for this program."
      write, "This error was found by: read an eof list."
  end if

End Subalgorithm
```

ALGORITHM EXAMPLE 8–14

In the while heading, both conditions must be true for the loop to continue. The check, (COUNT $<$ MAXINDEX), must be "strictly less than" rather than "less than or equal". Consider what happens if COUNT is equal to MAXINDEX, as it would be in the final pass through the loop with "less than or equal": COUNT would be incremented two lines later becoming MAXINDEX + 1, and would then be used to subscript the array—an error.

In similar fashion, we can construct a subalgorithm for a sentinel value loop:

```
Subalgorithm to read a sentinel data list filling in an array
          returning array X and its cursor COUNT.

  set COUNT to 0
  read, value
  while (COUNT < MAXINDEX) and (value ≠ sentinel) do
    increment COUNT by 1
    set X_COUNT to value
    read, value
  end while

  if value ≠ sentinel
    then
      write, "The data list is too long for this program."
      write, "This error was found by: read a sentinel list."
  end if

End Subalgorithm
```

ALGORITHM EXAMPLE 8–15

and also for a header value loop:

```
Subalgorithm to read a header data list filling in an array
          returning array X and its cursor COUNT.

  read, header
  if header > MAXINDEX
    then
      write, "The data list is too long for this program."
      write, "Only the first ", MAXINDEX,
                       "values will be read."
      write, "This error was found by: read a header list."
      set header to MAXINDEX
  end if
```

```
    for COUNT ← 1 to header do
        read, value
        set X_COUNT to value
    end for

    set COUNT to header

End Subalgorithm
```

ALGORITHM EXAMPLE 8–16

8.4.2. Summing the Values in an Array

Now that we have subalgorithms to read in any kind of data list, we can use them as higher-level statements in constructing solutions to more complex problems. Similarly, we must specify as subalgorithms the other tools we need to construct more modular solutions. These other tools, summing and averaging lists and finding the smallest and the largest value in lists, can also be obtained by properly modifying our previous solutions.

One of the major differences between the way we wrote these modules before and the way we write them as subalgorithms is that *we assume the data has already been read in and put into an array and a placekeeper variable set* so that we find the "end" of the data. As we develop these modules, we obtain the values we work with from the array and do not read in new data.

Since the array is already filled and the placekeeper variable is set, the simplest way to construct the new subalgorithms is to note that looking at each value in the array is much like reading in a header value data list. We *will not read* data, however; we will only *examine* the elements of the array we are given, one by one from the first. The summing algorithm for a header value data list is

```
Start Algorithm

    set sum to 0
    read, header
    for i ← 1 to header do
        read, value
        set sum to sum + value
    end for
    write, "The sum of the data list is: ", sum
    stop

End Algorithm
```

ALGORITHM EXAMPLE 8–17

The first thing we must do is restate this as a subalgorithm. The preconditions for this summing subalgorithm should be the array of values to be summed and the number of values in the array (equal to the value of the placekeeper variable). The postcondition is the sum of the values in the array. We need not change either the values in the array or the placekeeper variable; so these are not postconditions. Changing this much, we get

```
Subalgorithm to sum the values in an array
             given array X and its placekeeper N, and
             returning sum.

  set sum to 0
  read, header
  for i ← 1 to header do
    read, value
    set sum to sum + value
  end for
  write, "The sum of the data list is: ", sum

End Subalgorithm
```

ALGORITHM EXAMPLE 8–18

The next modification is to *remove* the read statements. They no longer have any purpose, since we assume that the data has already been read and put into the array that is given to us as a precondition.

```
Subalgorithm to sum the values in an array
             given array X and its placekeeper N, and
             returning sum.

  set sum to 0
  for i ← 1 to header do
    set sum to sum + value
  end for
  write, "The sum of the data list is: ", sum

End Subalgorithm
```

ALGORITHM EXAMPLE 8–19

Now we have to decide what to do with the variable header. It was used previously to specify the number of data items to read, but we needn't read anything in this subalgorithm. Instead of header, we use N. In the subalgorithm heading, we called the array's placekeeper variable N, which corresponds to what header stood for: the number of values to be summed.

We must also decide what to do with the variable value—it is no longer defined. In the old algorithm, value was used to hold the latest value obtained from the input list. In this situation, we replace value by the next item to be added into the sum—in this case, the ith element of the array. We replace the variable value by the array reference X_i.

Further, we don't need the write statement anymore. Presumably, the algorithm that calls this summing subalgorithm will obtain the sum via the parameter list as a postcondition and use it as it sees fit, including writing it out (if necessary).

This is the first time we've had this perspective on subalgorithms: we are writing a subalgorithm that will be used by another algorithm whose purpose we do not know. In a situation like this, our job is to do what we're expected to do, and not complicate matters for the calling algorithm. We shouldn't write anything out because the calling algorithm might need only the sum for an intermediate computation and doesn't want (or need) the sum written out. Our job is to add up the values in the data list, that's all.

One exception to this rule is that we *should* write out messages when we detect an error—that is, an impossible circumstance, like being asked to divide by zero, find the square root of a negative number, etc.

The summing subalgorithm is now written this way:

```
Subalgorithm to sum the values in an array
              given array X and its placekeeper N, and
              returning sum.

  set sum to 0
  for i ← 1 to N do
     set sum to sum + Xi
  end for

End Subalgorithm
```

ALGORITHM EXAMPLE 8–20

There is one small snag. Looking back at the solution we have just completed, we should ask what happens when there are no values in the array, that is, when N = 0. We return a sum of zero. That's fine—except that, technically, there is no sum. A sum of zero exists, like $-1 + 1$, but if we add up no values at all, a sum does not exist. What do we do about it?

It depends. We could assume that the calling algorithm will catch the error, if it is important in the context of the bigger problem it is solving, or we could write a message saying that the sum of zero values cannot be computed (the "That does not compute" message). For now, we choose the first option and assume that the calling algorithm will check to be sure there is at least one value in the array (if it matters). We leave the decision as to whether an error has occurred to the highest-level caller that can detect the error. We certainly do not halt the entire program because we do not know (in the subalgorithm) the complete context.

8.4.3. Averaging the Values in an Array

Now we can tackle the averaging problem. Fortunately, our job is made much easier by the fact that we've just completed the summing subalgorithm and can use it in our solution. (Never reinvent the wheel.) Using the summing subalgorithm greatly simplifies our task.

```
Subalgorithm to average the values in an array
          given the array LIST and its placekeeper N, and
          returning AVERAGE.

  if N > 0
    then
      sum the values in an array
        given array LIST and its placekeeper N,
        and returning TOTAL.
      set AVERAGE to TOTAL/N
    else
      set AVERAGE to 0.0
      write, "Error: an attempt was made to average an empty list."
      write, "This was detected in subalgorithm average the values"
      write, "in an array."
  end if

End Subalgorithm
```

ALGORITHM EXAMPLE 8–21

The first task in computing the average is to compute the sum, which we do by using our summing subalgorithm. The next task is to divide the sum by the number of values in the sum. If the number of values is zero, we cannot divide. Therefore we must check for this and for the possibility of a negative (absurd) count as well. If we find an error of this sort, the messages we write should make sense to a nonprogrammer, but give a programmer enough information to correct the problem. If there is no error, we simply compute the sum, do the division, and stop.

Here we have called a subalgorithm from within another subalgorithm. There is no problem in doing this, and indeed many reasons why we should. We needed to sum the values in an array, and we already have a subalgorithm to do this. It would be silly (and more work) to rewrite the summing subalgorithm as part of the averaging subalgorithm.

Another reason for calling a standard subalgorithm whenever we can is that if someone finds a better way to do one of them and rewrites only the standard subalgorithm, all (sub)algorithms that call it benefit from the better way of doing it.

Finally, calling a subalgorithm reduces the conceptual complexity of a problem's solution by reducing the number of details (preconditions and postconditions) we must keep track of.

8.4.4. Finding the Smallest or Largest

The subalgorithm we will develop in this section finds the smallest or the largest value in an array. Again, the easiest approach is to start with a similar problem and modify it to our purposes—and we have this algorithm from Chapter 3 to find the smallest:

```
Start Algorithm

   read, champion

   while there is data left do
      read, challenger
      if challenger < champion
         then
            set champion to challenger
      end if
   end while

   write, "The smallest value in the data list is: ", champion
   stop

End Algorithm
```

ALGORITHM EXAMPLE 8–22

Let's change it to a subalgorithm and delete the write statement. The preconditions will again be the same as for the summing and averaging subalgorithms: an array that already contains values and a placekeeper variable whose value tells us how many elements are stored in the array. The postcondition is the smallest value in the array. In the subalgorithm, we will call the array values and its placekeeper size. We will also call the "champion" by the name small (to better remember it).

```
Subalgorithm to find the smallest value in an array
            given the array values and its placekeeper size,
            and returning small.

  read, small

  while there is data left do
    read, challenger
    if challenger < small
      then
        set small to challenger
    end if
  end while

End Subalgorithm
```

ALGORITHM EXAMPLE 8–23

As before, the values we will use are already in the array and should not be read in. We can eliminate the reads and replace them everywhere by references to the array, but which element of the array? The first read statement is supposed to "prime" small with the first data value. That read should then be replaced by set small to $values_1$. The read inside the loop is supposed to obtain the rest of the values, one for each pass through the loop. After obtaining a particular value, the if statement is to compare that value with small. It would appear that while there is data left do is inappropriate for two reasons. Because we are not reading data, to check whether data exists is inappropriate. Also, we need an index that starts at 2 and proceeds through size, the sequence of indexes that allows us to examine each element of the array after the first. We should use a for loop here.

```
Subalgorithm to find the smallest value in an array
            given the array values and its placekeeper size,
            and returning small.

  set small to values₁

  for k ← 2 to size do
    if valuesₖ < small
      then
        set small to valuesₖ
```

```
      end if
    end for

End Subalgorithm
```

ALGORITHM EXAMPLE 8–24

Finally, we have it—or do we? There is *always* a snag, and this is the same one as for the summing and averaging subalgorithms. What are we to do when size is zero—when there are no values and so no smallest value? Currently, our subalgorithm will return the undefined contents of $values_1$ as small since the for loop will not be executed. We have a dilemma: Do we write an error message or not?

In this case, let's check for the error condition, but write no error message. The check consists of enclosing the entire body in an if statement that checks the condition size > 0. We set the subalgorithm up so it will be easy to add an error message if we need it later; but we do nothing about it now, and assume (as we did before) that the calling algorithm should check for the error.

```
Subalgorithm to find the smallest value in an array
        given the array values and its placekeeper size,
        and returning small.

  if size > 0
    then
      set small to values₁

      for k ← 2 to size do
        if valuesₖ < small
          then
            set small to valuesₖ
        end if
      end for
    else
      {add error messages here if it matters that an}
      {undefined value is returned as small when there}
      {are no values in the array}

  end if

End Subalgorithm
```

ALGORITHM EXAMPLE 8–25

To make Algorithm 8–25 find the largest (rather than the smallest) value in the array, simply change

```
if values_k < small
```

to

```
if values_k > small
```

It's also appropriate to change the name small to large and find the smallest to find the largest so as not to confuse anyone.

```
Subalgorithm to find the largest value in an array
        given the array values and its placekeeper size,
        and returning large.

  if size > 0
    then
      set large to values_1

      for k ← 2 to size do
        if values_k > large
          then
            set large to values_k
        end if
      end for
    else
      {add error messages here if it matters that an}
      {undefined value is returned as large when there}
      {are no values in the array}
  end if

End Subalgorithm
```

ALGORITHM EXAMPLE 8–26

8.5. RECORDS AND THEIR USES

A **record** is a collection of memory cells, each of which can hold a different data type and a set of field names (to allow us to refer to each element). We could write algorithms without records; in fact some programming languages (like

FORTRAN and BASIC) do not allow the use of records. We use records to group information about an object or entity (as opposed to keeping separate variables that contain the same information as that in a record, but implicitly grouped by the way we use the variables.) It is always better to be explicit in an algorithm whenever possible. The more explicit the algorithm the better our chance of understanding it and making it work properly.

Suppose—as an example of how to use records in an algorithm—we have a record variable named "volume" of type book record that has the same field names as before:

Title	(string of characters)
Author	(string of characters)
Publisher	(string of characters)
DateOfPublication	(string of characters)
NumberOfPages	(integer)
Hardback	(true or false)
ISBN	(string of characters)
LibraryOfCongressRef	(string of characters)
PurchasePrice	(real)

FIGURE 8–4 Book Record Definition

One advantage of grouping these nine items of information together is that we can refer to the whole collection by the record variable name, rather than use nine separate variable names. As usual, we state a subalgorithm to read in such a record.

```
Subalgorithm to read in a book record
                returning the book record, volume.

  read, volume field Title
  read, volume field Author
  read, volume field Publisher
  read, volume field DateOfPublication
  read, volume field LibraryOfCongressRef
  read, volume field PurchasePrice
  read, volume field ISBN
  read, volume field Hardback
  read, volume field NumberOfPages

End Subalgorithm
```

ALGORITHM EXAMPLE 8–27

In this example, the data is arranged in the order we read it in. This does not correspond to the order of the fields within the book record, but it does not matter. When we read in the data, we specify which field we wish to use, as we did when we used a simple variable. We treat each field individually when we need to change or examine that field's content.

8.6. COMPOSITION OF TYPES

To this point, we have discussed only the possibility of using simple types as the elements of arrays and fields of records. However, when we have more complex objects to represent, it is convenient to allow the elements of an array to be records or other arrays, and the fields of records to be arrays and other records. If we were dealing with a small library of 500 books, we could represent such a library as an array whose indexing sequence is 1 to 500 and whose element type is a book record. We could read the entire library data base into such an array using a simple modification of the read-an-eof-data-list subalgorithm we developed earlier in this chapter:

```
Subalgorithm to read an eof data list filling in an array
          returning array X and its cursor COUNT.

  set COUNT to 0
  while (COUNT < MAXINDEX) and (there is data left) do
    read, value
    increment COUNT by 1
    set X_COUNT to value
  end while

  if there is data left
    then
      write, "The data list is too long for this program."
      write, "This error was found by: read an eof list."
  end if

End Subalgorithm
```

ALGORITHM EXAMPLE 8–28

The only changes we need to make to this subalgorithm is to update the read statement, assignment, and subalgorithm heading to reflect the fact that we are now reading in an entire book record instead of a simple data value. We have defined the read instruction so that it can read only simple types. Fortunately, we already have a subalgorithm to read in a single book record from the last section.

We extend our definition of assignment to assignment of array and record variables, regardless of the number of elements or fields. (Some programming languages allow this, such as Pascal and COBOL. Others, such as FORTRAN and BASIC, do not.) The updated version of the subalgorithm is then

```
Subalgorithm to read an eof data list of book records
          returning array X of book records
          and its cursor COUNT.

  set COUNT to 0
  while (COUNT < MAXINDEX) and (there is data left) do
    read in a book record receiving the book record, value
    increment COUNT by 1
    set X_COUNT to value
  end while

  if there is data left
    then
      write, "The data list is too long for this program."
      write, "This error was found by: read an eof list."
  end if

End Subalgorithm
```

ALGORITHM EXAMPLE 8–29

8.7. SUMMARY

This chapter discussed the composite data types *array* and *record*. We first discussed the problems that require more than variables of simple types for their solution. The example was the reverse-a-list problem. We showed a solution that uses only variables of simple types, and that solution soon became too long when a list of a reasonable length was to be reversed.

In conjunction with arrays, we described the need for a *cursor* or *placekeeper* variable to keep track of the last index value used. We also expanded our definition to include the concept of an array as a composition of an *index type* and an *element type*. We described the index type as a contiguous sequence of values from a simple, countable type (integer, character, or boolean).

Several standard subalgorithms were derived from previously defined modules to build a set of tools for writing more complex algorithms. The subalgorithms in this chapter are: reading data into an array (using sentinel, end-of-file, and header-type data lists); finding the sum of values in an array; finding the average of the values in an array, and finding the smallest/largest value in an array.

We showed you how to read values into the *fields* of a record, and discussed the use of records in organizing data to be stored in memory.

8.8. EXERCISES

8.01. Write an algorithm to help an apartment locator service match people and apartments.

Each apartment is described in terms of 30 possible features. A 'Y' would indicate it has the feature and an 'N' that it does not.

A customer fills out a form indicating which of the 30 features she/he wants in an apartment. A 'Y' means that the feature is desired and an 'N' means it is not.

The data file contains several customer requests. Each request consists of the 30 Y/N responses corresponding to features, the customer's name, and, on the next line, a number indicating how many apartment descriptions with addresses follow. The names and addresses contain no more than 20 characters, and are separated from the features by a single blank space.

The algorithm must write the customer's name, each address, the number of features matched, and (for each address) either 'POSSIBLE' or 'NO'. 'POSSIBLE' is printed if the number of matches is at least 24, and 'NO' is printed if the number of matches is less than 24.

Sample Input

```
YYYYYNYYYYYNYYYYYNYYYYYNYYYYYN Ronald Reagan
2
YYYYYNYYYYYNYYYYYNYYYYYNYYYYYY 1600 Pennsylvania
YYYYYYYYYYYNYYYYYNYYYYYNYYYYYN Hollywood and Vine
YNYNYNYNYNYNYNYNYNYNYYYYYYYYYY Jay Hawk
3
NYNYNYNYNYNYNYNYNYNYYYYYYYYYYY 2300 Kasold
YNYNNYNYNYNYNYNYNYNYYNYYNYNYNYNY 1501 E. 15th
YNYNYNYNYNYNYNNNNNYYYYNYNYYYYN 701 New Hampshire
```

Sample Output

```
Ronald Reagan
1600 Pennsylvania      29 matches      POSSIBLE
Hollywood and Vine      29 matches      POSSIBLE
```

Jay Hawk		
2300 Kasold	9 matches	NO
1501 E. 15th	12 matches	NO
701 New Hampshire	24 matches	POSSIBLE

8.02. A *palindrome* is a character string that reads identically in both directions, for example:

NOW BOB WON

A *packed palindrome* is a character string that reads identically in both directions after all nonalphabetic characters are removed, for example:

MADAM, I'M ADAM.

Each line of the data file contains a character string.

Write an algorithm that reads in one character string at a time, echo-prints the character string, and reports whether it is a palindrome, a packed palindrome, or neither. Read until you reach the end of the data file. After processing all the character strings, print out the total number of each of the three types of strings.

Sample Input

```
ABLE WAS I ERE I SAW ELBA
A MAN, A PLAN, A CANAL — PANAMA
HUNGRY WAS I ERE I ATE DINNER
```

Sample Output

```
ABLE WAS I ERE I SAW ELBA
    is a palindrome.
A MAN, A PLAN, A CANAL — PANAMA
    is a packed palindrome.
HUNGRY WAS I ERE I ATE DINNER
    is not a palindrome.
There were 1 palindrome(s).
There were 1 packed palindrome(s).
There were 1 string(s) that were not palindromes.
```

8.03. The U.S. team wants to select runners to compete in the Olympics, and to qualify for the team, each contestant must have run a race in less than the qualifying time. You are given the results (finishing times) of an unknown number of trials for various races. For each trial, the input data is arranged as follows:

Line 1 has the number of contestants for this trial.
Line 2 has the list of finishing times for this trial, where the times are ordered according to the contestant number (e.g., the first time in the list corresponds to contestant 1).
Line 3 has the qualifying time for this race.

The end of the data file is indicated by 0 for the number of contestants. For each trial, do the following:

1. Print out the qualifying time for this race.
2. Print out the number of each contestant who qualified (i.e., had a time less than or equal to the qualifying time).
3. Print out the number of contestants and the number who qualified.

Sample Input

```
7
2.5 2.3 2.1 1.8 1.7 2.7 2.3
2.0
9
2.7 2.9 2.4 3.0 1.7 2.5 1.6 2.3 1.8
1.0
0
```

Sample Output

```
In race 1 the qualifying time was 2.0
   Contestant 4 qualified with a time of 1.8
   Contestant 5 qualified with a time of 1.7
2 contestants qualified out of 7 contestants.

In race 2 the qualifying time was 1.0
None of the 9 contestants qualified.
```

8.04. The data file is a student record file, and each line in the file has the following information:

1. Student name—last name followed by first name; each, at most, 15 characters
2. Student ID—integer
3. Classification—integer
4. School—1 character
5. Score on exam 1—integer
6. Score on exam 2—integer
7. Score on final exam—integer

Sample Input

```
Donne John
123456 3 J
80 72 80
Milton John
111222 1 A
70 60 74
Ofarc Joan
111111 4 A
90 81 98
```

Read until the end of the data file; then, compute the final letter grade for all students and print the student information, along with their final letter grades. The letter grade is determined by the weighted sum of their scores on the tests:

```
Exam 1     = 30% of grade
Exam 2     = 30% of grade
Final exam = 40% of grade
```

The weighted sum is then compared to the weighted sum of the averages of each test. If "MiddleC" is the sum of the averages, the grade is determined by this table:

Student Total	Letter Grade
≥ MiddleC +15	A
≥ MiddleC + 5	B
≥ MiddleC − 5	C
≥ MiddleC −15	D
otherwise	F

This is a standard curve, whose average is the middle of the "C" range. However, to avoid penalizing students if they do well on a test, use an average of 75 for that test (which would yield a 90-80-70-60 curve) if the average is higher than 75.

Your output should include the information given in the sample output below.

Sample Output

The average on the first test was 80.0 (will use 75.0)
The average on the second test was 71.0 (will use 71.0)
The average on the final exam was 84.0 (will use 75.0)
The curve was based on a weighted average of 73.8

Student Name	ID	Class	School	Total	Grade
Donne John	123456	3	J	77.6	C
Milton John	111222	1	A	68.6	D
Ofarc Joan	111111	4	A	90.5	A

8.05. This problem, which deals with arrays and searching (in a rudimentary sense), is perennial for CS instructors: How to assign grades? (It won't be as bad for you as it is for me, because I'm making it simple.)

As you read in a list of names and test scores (1 test score per student), prepare to find their average. After all names and scores have been read, print each student's name, test score, and an evaluation "Satisfactory", "Unsatisfactory", or "Outstanding". The criteria on which evaluation is based are:

Outstanding	scores higher than the average + 10
Satisfactory	scores between average − 10 and average + 10
Unsatisfactory	scores lower than average − 10

Assume that no student's name is longer than 25 characters. Also, you should be sure that test scores are in the range 0-100. When you read the data in, echo-print the data with any error messages you deem necessary. You should also print out the average as soon as you compute it (i.e., after you've read all the data). Finally, print the name, score, and evaluation for each student (under suitable headings).

8.06. The data file contains a series of words, arranged 1 per line. Write an algorithm that produces a histogram for the frequencies of each word in the file. (A histogram is like a bar graph and consists of a row of asterisks ["*"] that show the relative frequencies of each word, where each row is labeled with the word concerned.) The histogram should be scaled (by the algorithm) so that the longest row has 20 asterisks. You may assume that no word is longer than 12 characters and that there are no more than 100 distinct words (although each word may occur any number of times).

Sample Input

this
is
the
input
data
this
data
is
this

Sample Output

this 3 ********************
is 2 *************
the 1 *******
input 1 *******
data 2 *************

8.07. Input data consists of information for a student register, a course register, and an enrollment register:

STUDENT REGISTER (information for each student of the university)
Student's name: up to 20 characters
Student's ID number: integer
Student's class: 1 = freshman, 2 = sophomore, 3 = junior, 4 = senior

COURSE REGISTER (information for each course offered)
Course number: up to 5 characters
Course name: up to 20 characters

ENROLLMENT REGISTER (showing which students enrolled in which courses)
Course number: up to 5 characters
Student number: integer

The first line of the data file contains the number of students, the number of courses, and the number of enrollments. The remaining lines contain the student, course, and enrollment register lines, in that order. When reading the data for the registers, you may assume that all input is valid, and there are no more than 25 enrollment lines, 10 student lines, and 10 course lines.

Processing consists of reading the information for the registers and producing the listings described below. To minimize the need for error-checking searches, you may assume—for every student (or course) number in the enrollment register—that there is information for that student (or course) in the appropriate register.

Listings to be Printed

1. Registers. Separately, print a complete Student Register, complete Course Register, and complete Enrollment Register, with identifying headings.
2. Course Rosters. Rosters should be printed for each course offered by the university. Include course number and name, followed by the number and name of all students enrolled in a course. The list of student names should be followed by the number of students enrolled in the course.

8.08. Write an algorithm to create and maintain a data base for Jayhawk National's bank-card department. Your algorithm must keep track of the accounts for all bank-card customers, including their transactions.

Your algorithm will read in customer information (a 6-digit account number, checking account balance, and savings account balance) until an account number of zero is encountered.

This information will be followed by a list of transactions for the above accounts (each transaction will have a 'C' or 'D' to indicate credit or debit, a 'C' or 'S' to indicate checking or savings, a 6-digit account number, and the amount of the transaction). A *credit* means you add money to the account; a *debit* means you subtract money from the account. You cannot debit more money than is in the account.

Your algorithm must check for the following errors in the transactions:

a. Invalid transaction code (not a 'C' or 'D')
b. Invalid account type (not a 'C' or 'S')
c. Invalid account number
d. Attempts to overdraw an account

Your output is a list of all transactions and appropriate messages for errors in the transactions. Final account balances for all customers should be printed out in table form.

Customer information lines will look like this:
XXXXXX YYYY.YY ZZZZ.ZZ
where XXXXXX is the account number
YYYY.YY is the checking account balance
ZZZZ.ZZ is the savings account balance

Transaction lines will look like:
T A XXXXXX DDDD.DD
where T is the type of transaction (either 'C' or 'D')
A is the type of account (either 'C' or 'S')
XXXXXX is the account number
DDDD.DD is the amount of the transaction

Sample Input

```
125643    1000.00     23.65
764532      29.00    521.97
000000     234.21      1.00
C C    764532     34.98
T S    125643    100.00
D S    125643     20.00
C C    122345     11.11
```

Sample Output

```
Account number 764352: Checking deposit of $ 34.98
Account number 125643: Illegal transaction code - T
Account number 125643: Savings withdrawal of $ 20.00
Account number 122345: No such account

              Final Balances

Account #      Checking      Savings
-------------------------------------------------------------------
  125643        1000.00         3.65
  764532          63.98       521.97
```

9
Algorithmic Modules That Use Composite Data Types

In this chapter we will discuss three important modules that require composite data types: searching a list for a given value, building a list of unique values from data, and sorting a list of values. The major data type we will use is the array. Initially, we will consider only an array whose element type is simple, but later we will consider arrays whose element type is a record.

9.1. SEARCHING

Searching determines if a particular value (held in a simple variable) occurs at least once in a list of values (presumably in an array). If it does, we might also be interested to know the position where the first match occurs. For example, if we were given the list 'a', 'b', 'd', 'm', 'd', and 'x', and asked if 'd' occurs, the answer is yes, and it occurs in position 3. (It also occurs in position 5, but we were not asked to check for that.)

To find the answer, we start at the front of the list and compare the value we are asked to find with each element of the list, successively. As we proceed from one element to the next, we count the elements, so that if we find a match we are able to say where it occurs. Once we find a match, we don't look further (unless your curiosity gets the better of you).

If we mimic this in algorithmic terms, we must specify things more carefully. The preconditions for a subalgorithm to search through a list are (1) the list (an array), (2) the array's cursor or placekeeper (the array and its cursor always go together), and (3) the value we are to search for. The postconditions for such a subalgorithm are (1) an indication whether we found a match (true or false) and (2) if we found a match, in what position we found it. This last postcondition is a new wrinkle. It is given a value only if the first postcondition is true; otherwise it is undefined.

Starting at element 1, we step through the array and at each step compare the value we are looking for with an element of the array. As soon as we find a match, we stop. If we don't find a match, we compare the value with every element of the array. After we have done this, and found no match, we conclude that no element in the array matches the value. This is called a **linear search**.

As we start the subalgorithm, we have not found the value we are looking for; so we start the value of a boolean variable found at false. If we find the value while we search the array, we change the value of found to true.

At the start of the algorithm, we must start an index variable at zero to indicate that we have not yet compared the value for which we are searching with any value in the array. Then, to "look at" successive values in the array, we increment this index and look at the content of the array at this index position. We know we must stop looking in the array when the index value reaches the value of the placekeeper, the position of the last defined element in the array. This is called **scanning an array**.

We have done this before, when we summed the values in an array. In that case, however, we had to scan the entire defined contents of the array, i.e., from index position 1 through the index position in the placekeeper. The obvious looping instruction in that situation was a for loop. In searching an array, however, we scan only until we find a match (or exhaust all the values in the array).

Here, we use a while loop with two conditions: one that ensures that the scan index is less than the placekeeper and one that stops the loop as soon as the value of the variable found becomes true.

One way to write the standard scanning module is

```
[initialization]
set scan to 0
while scan < PlaceKeeper do
   increment scan
   [process list_scan]
end while
[post loop processing]
```

ALGORITHM EXAMPLE 9–1

To change this standard scanning module into the searching subalgorithm, we add a subalgorithm heading:

```
Subalgorithm to find a value in an array,
        given array X, its PlaceKeeper,
              and the value to find, Y,
        returning found and position.

  [initialization]
  set scan to 0
  while scan < PlaceKeeper do
    increment scan
    [process list_scan]
  end while
  [post loop processing]

End Subalgorithm
```

ALGORITHM EXAMPLE 9–2

Next, we specify the [initialization] part and modify the while loop termination condition to use the boolean variable found (discussed earlier):

```
Subalgorithm to find a value in an array,
        given array X, its PlaceKeeper,
              and the value to find, Y,
        returning found and position.

  set found to false
  set scan to 0
  while (scan < PlaceKeeper) and (not found) do
    increment scan
    [process list_scan]
  end while
  [post loop processing]

End Subalgorithm
```

ALGORITHM EXAMPLE 9–3

We are left with two sections to refine: [process $list_{scan}$], and [post loop processing]. In order to [process $list_{scan}$], we compare Y with X_{scan}, and if they are equal, we set found to true. If they are not equal, we do nothing.

```
Subalgorithm to find a value in an array,
        given array X, its PlaceKeeper,
              and the value to find, Y,
        returning found and position.
```

```
set found to false
set scan to 0
while (scan < PlaceKeeper) and (not found) do
  increment scan
  if X_scan = Y
    then
      set found to true
  end if
end while
[post loop processing]

End Subalgorithm
```

ALGORITHM EXAMPLE 9–4

The [post loop processing] section for the search algorithm copies the value in scan into the postcondition position. Making this modification, we obtain the final search subalgorithm:

```
Subalgorithm to find a value in an array,
        given array X, its PlaceKeeper,
              and the value to find, Y,
        returning found and position.

set found to false
set scan to 0
while (scan < PlaceKeeper) and (not found) do
  increment scan
  if X_scan = Y
    then
      set found to true
  end if
end while
set position to scan

End Subalgorithm
```

ALGORITHM EXAMPLE 9–5

In this subalgorithm we assume that X is an array with an index type of 1, 2, 3, . . . , MAXINDEX, and that the element type of X is appropriate to the problem we need to solve. PlaceKeeper should be an integer, $0 \leq$ PlaceKeeper $\leq$ MAXINDEX. Y should have the same type as the element type of X. The postcondition found is a boolean parameter and position is an integer with $0 \leq$ position $\leq$ MAXINDEX.

9.2. BUILDING LISTS WITH UNIQUE ENTRIES

Many problems that we will be asked to solve involve reading in a list of data values and counting the number of times each unique value occurs. There are many ways to do this, but an efficient method uses the search subalgorithm we developed in the last section. The basic idea is that when we read a value, it is either a new value or it has already been seen at least once. If it's a new value, we add it at the bottom of the list of unique values we are building and remember that we have seen it once. If we've seen this value before, we increment the count we are keeping for that value. This implies that either we have two arrays, one for the values and one for the count of the times we have seen each value, or we have an array whose elements are records that contain a value and a count.

Initially, we use two arrays. Keeping several arrays to hold information about a single data list is called keeping **parallel arrays**. (They are called *parallel* because the information in the first position in each array is about the same item in the input list; the second element in each array is about another single item from the input list; and so on.) It is preferable, when possible, to use a single array whose elements are records, with one record holding all the information about one data item. In some languages, however, records are not defined; so this is not possible (BASIC and FORTRAN are the most widely used languages that do not support the concept of a record).

For example, if the data list is

```
1
2
5
2
6
5
1
7
```

we should produce as output:

```
The unique values in the data and their frequency
Value     Frequency
  1           2
  2           2
  5           2
  6           1
  7           1
```

Here, we use two arrays. The first contains the unique values, and the second contains the frequency (count of the number of times) the corresponding value was seen in the input list.

In general, this problem can easily be solved if we take advantage of the searching subalgorithm we already have. The general form is based on the standard read-an-eof-data-list-into-an-array module from the last chapter. In this example, size will be the placekeeper for the array that contains the values as we store them into the array. Size will also be the placekeeper for the array that contains the frequencies. Since both arrays are parallel, they contain information in exactly the same element positions, and thus one placekeeper can serve for both arrays.

Below, we reproduce the subalgorithm to read an end-of-file data list into an array.

```
Subalgorithm to read an eof data list filling in an array
        returning array X and its cursor COUNT.

  set COUNT to 0
  while (COUNT < MAXINDEX) and (there is data left) do
    read, value
    increment COUNT by 1
    set X_COUNT to value
  end while

  if there is data left
    then
      write, "The data list is too long for this program."
      write, "This error was found by: read an eof list."
  end if

End Subalgorithm
```

ALGORITHM EXAMPLE 9–6

The essential difference between reading the entire data list into array X and what we need is that we store the value just read *only if it is not already in the array*. We also need to count the occurrences of each distinct value, but this is easy once we resolve the essential difference. First, let's change the subalgorithm heading and the names of the variables X and COUNT so that they are more meaningful in the context of this problem.

```
Subalgorithm to construct a list of unique values from an eof data list
        returning array unique and its cursor size.

  set size to 0
  while (size < MAXINDEX) and (there is data left) do
```

```
    read, value
    increment size by 1
    set unique_size to value
  end while

  if there is data left
    then
      write, "The data list is too long for this program."
      write, "This error was found by: construct unique list."
  end if

End Subalgorithm
```

ALGORITHM EXAMPLE 9–7

The two statements in Algorithm Example 9–7 that control the storing of values into the array are the second and third lines in the while loop. Rather than store the data values unconditionally, we should search the array unique of length size to see if the value just read is already there. If it is, we should increment the corresponding position in array frequency by 1 to indicate we have seen the value one more time. If it is not found, *then* we need to add it to the array unique, with the corresponding position in array frequency set to 1 to indicate we have seen the value just once.

```
Subalgorithm to construct a list of unique values from an eof data list
        returning arrays unique and frequency and their cursor size.

  set size to 0
  while (size < MAXINDEX) and (there is data left) do
    read, value
    if value is not found in the array unique
      then
        increment size by 1
        set unique_size to value
        set frequency_size to 1
      else
        increment frequency_position by 1
    end if
  end while

  if there is data left
    then
      write, "The data list is too long for this program."
```

```
    write, "This error was found by: construct unique list."
  end if

End Subalgorithm
```

ALGORITHM EXAMPLE 9–8

(We also added the array frequency as a postcondition in the subalgorithm heading.)

The only problem with this subalgorithm is that the condition "if value has not been seen before" is a little vague. We can make it precise, however, if we call the subalgorithm we wrote to search for a value. Given the list of unique entries (which starts out empty) and a value to look for, the subalgorithm will return a boolean value of true, if the value is already in the list, or false, if the value is not in the list. The search subalgorithm also returns an index value if it finds the search value in the list, indicating the element number at which the match was found.

All we do to make the algorithm build a unique list is call the search subalgorithm before the if statement and use the boolean parameter it returns as the check for "if the value has not been seen before." We also write out the values and frequencies to conform to the output shown at the beginning of this section.

```
Subalgorithm to construct a list of unique values from an eof data list
        returning arrays unique and frequency and their cursor size.

  set size to 0
  while (size < MAXINDEX) and (there is data left) do
    read, value
    find a value in an array, given array unique,
                                its placekeeper size, and
                                the value to find, value,
                         returning found and position.
    if not found
      then
        increment size by 1
        set unique_size to value
        set frequency_size to 1
      else
        increment frequency_position by 1
    end if
  end while

  if there is data left
    then
```

```
        write, "The data list is too long for this program."
        write, "This error was found by: construct unique list."
    end if

    write, "The unique values in the data and their frequency"
    write a blank line
    write, "     value      frequency"
    write a blank line

    for i ← 1 to size do
      write, unique_i, frequency_i
    end for

End Subalgorithm
```

ALGORITHM EXAMPLE 9–9

A better way of building a list of unique data values with associated frequency counts is to use a single array whose element type is a record. In the previous example, we used two parallel arrays—one to hold the actual data values and the other to hold the frequency counts. Each of these had simple element types. We now propose to use a single array whose element type is a record that consists of two fields: a data field and a frequency field.

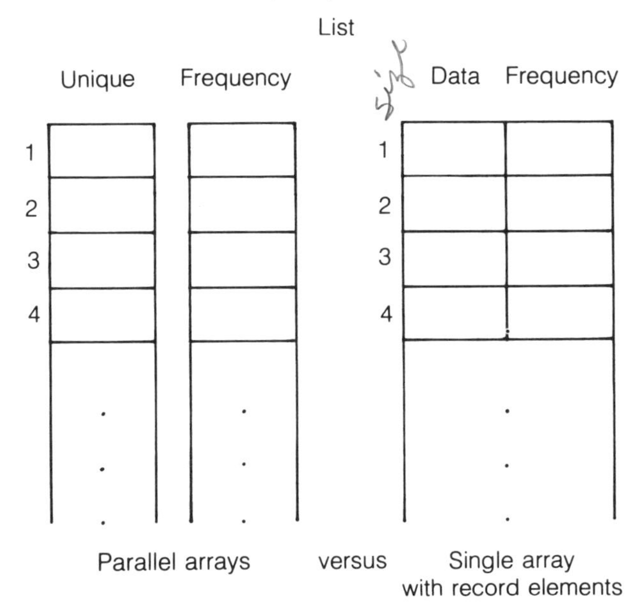

FIGURE 9–1 Parallel Arrays vs. Single Array of Records

When we use parallel arrays, we use the *convention* that corresponding elements in each array are associated with one another, i.e., contain information about the same data item. Except by tracing the algorithm and determining that the arrays are used in a parallel fashion, however, *there is no explicit connection between the two arrays*. This is poor programming if a better alternative is available.

When a single array with record elements is used, the record structure groups each piece of information about the same data item so there is no doubt that the pieces are associated with one another. The one array, then, allows the grouping of several data items. In our previous example, each data item consisted of two pieces: a value read from the input list and the number of times that value occurred in the input list. We now declare a record type, called OneTime, with two fields: data and frequency, and an array, list, whose element type is OneTime.

To modify the subalgorithm that we used with parallel arrays for a single array with records, we replace each reference to one of the parallel arrays with the equivalent reference to a record field within the single array, list. The table below shows the corresponding equivalent references.

In Parallel Array Subalgorithm	*In Single Array with Records*
$unique_i$	$list_i$ field data
$frequency_i$	$list_i$ field frequency

By making appropriate substitutions into the parallel array subalgorithm, we obtain

Subalgorithm to construct a list of unique values from an eof data list
returning array list and its cursor size.

set size to 0
while (size < MAXINDEX) and (there is data left) do
  read, value
  find a value in an array, given array list,
    its placekeeper size, and
    the value to find, value,
    returning found and position.
  if not found
    then
      increment size by 1
      set $list_{size}$ field data to value
      set $list_{size}$ field frequency to 1
    else
      increment $list_{position}$ field frequency by 1
  end if
end while

if there is data left
  then
    write, "The data list is too long for this program."
    write, "This error was found by: construct unique list."
end if

write, "The unique values in the data and their frequency"
write a blank line
write, " value frequency"
write a blank line

for i ← 1 to size do
  write, $list_i$ field data, $list_i$ field frequency
end for

End Subalgorithm

ALGORITHM EXAMPLE 9–10

As usual, there is a snag. The search subalgorithm is written to search an array with a simple element type, not an array with a record type. Most programming languages do not allow the mixing of types this way; so we must modify the subalgorithm to search for a value in a manner similar to the calling subalgorithm. Fortunately, only two places refer to the array: the subalgorithm heading (which we can leave alone) and the comparison between an element of the array and the value we are searching for. We may leave the heading alone because we need not know the arrangement of the data within the array at that point. However, we must specify this in the if statement. In particular, we must refer to the field data within an element of X:

if X_{scan} field data = Y

The new subalgorithm to search for a value within a single array with record field data is

Subalgorithm to find a value in an array,
given array X, its placekeeper N,
and the value to find, Y,
returning found and position.

set found to false
set scan to 0
while (scan < N) and (not found) do
increment scan

```
        if X_scan field data = Y
          then
            set found to true
        end if
      end while
      set position to scan

    End Subalgorithm
```

ALGORITHM EXAMPLE 9–11

9.3. SORTING

Sorting data, a very common problem, can be the main problem (as in sorting zip codes for bulk mailings) or a subproblem in a larger problem (as in creating a book index). There are many ways to solve it, some much more efficient than others (see Chapter 10 for further discussion).

In essence, the sorting problem can be described quite simply. Given a list of values of a simple type, put them into ascending or descending order. We limit ourselves to simple types (at least here) because we know there is an ordering between any two elements of a simple type that allows us to determine what "ascending" and "descending" mean (i.e., whether one value is larger or smaller than another).

The method we use here is a **selection sort** (we *select* the next element from the list of possibilities). If we want to sort into ascending sequence, we take the original list and select the smallest element, which goes first in the sorted sequence. Again, select the smallest element from what is left, which is next in the sorted sequence. And so we continue until we select all the elements.

If we specify a subalgorithm to do this sorting, we need two preconditions and one postcondition. One of the preconditions is also the postcondition: the list of values. The other precondition is the cursor or placekeeper for the list. Since the number of elements in the list doesn't change, neither will the value of the cursor.

```
    Subalgorithm to sort an array into ascending order
            given array X and its cursor N, and
            returning the sorted array X.

      [sort the array]

    End Subalgorithm
```

ALGORITHM EXAMPLE 9–12

Starting values	Pass 1	2	3	4
>5	1	1	1	1
7	>7	3	3	3
3	3>	>7	3	3
3	3	3>	>7	5
1>	5	5	5>	>7

FIGURE 9–2 Tracing an Insertion Sort

For this plan, it seems we should use a for loop to scan a new array into which we put the selected values, one at a time; but there's a problem. If we use a separate array to store the part of the list we've already sorted, we have no way to tell which values in the original list we have used. We can't remove a value from an array (without a lot of shuffling of values), but we can exchange a value at one position for a value at another and (so to speak) sort the array into itself.

Figure 9–2 shows the content of a 5-element array during sorting. There is always one fewer pass than elements since the last element always ends up where it should be. The ">" on the left of a column shows the division point between the sorted (above the ">") and the unsorted part (at or below the ">"). On the right of the column, ">" shows the smallest element in the unsorted part. At each step, the smallest element is exchanged with the element at the top of the unsorted part of the list.

In the subalgorithm below, j is the index that determines the boundary between the sorted and unsorted parts of the array. Elements whose index values are less than j are already in the correct, sorted order. Elements whose index values are greater than or equal to j are yet to be sorted.

Subalgorithm to sort an array into ascending order
given array X and its cursor size, and
returning the sorted array X.

for j ← 1 to size-1 do

find the smallest value in the array with a starting index
given array X, its cursor size, the starting index j
and returning the position, p, of the smallest.
switch two variables, given X_j and X_p, and
returning X_j and X_p.

end for

End Subalgorithm

```
Subalgorithm to find the smallest value in the array
          with a starting index, given array X, its cursor N,
                                        the starting index K, and
                                    returning the position, p, of
                                        the smallest.

  set p to K
  for i ← K to N do
    if X_i < X_p
      then
        set p to i
    end if
  end for

End Subalgorithm

Subalgorithm to switch two variables given X and Y, and
          returning X and Y.

  set temp to X
  set X to Y
  set Y to temp

End Subalgorithm
```

ALGORITHM EXAMPLE 9–13

In this subalgorithm, we use two other subalgorithms. (We have seen the first one before: find the smallest value in an array.) We have not seen the other subalgorithm: switch two variables to exchange the values contained in the two variables X and Y. At first glance, we might think we can do this by using

```
set X to Y
set Y to X
```

but this doesn't work very well.

If the values of the variables before these two statements are X = 5 and Y = 3, their values afterward are X = 3 and Y = 3. Why? Trace the two-statement sequence.

The situation is analogous to two full glasses, a green one filled with milk and a blue one filled with 12-year-old scotch. How would you get the milk into the blue glass and the scotch into the green glass? Of course, you pour the scotch into an empty third glass, the milk into the now empty blue glass, and the scotch from the third glass into the now empty green glass. In the subalgorithm to switch the X and Y variables, X is like the green glass, Y is like the blue glass, and temp is like the empty third glass.

9.4. SUMMARY

This chapter presented three important problems that use arrays: searching an array, building a list of unique values from the input, and sorting a list.

The searching problem was discussed in terms of both parallel arrays and a single array with record elements. Subalgorithms were developed for both methods.

The subalgorithm to build a list of unique values from the input used the search subalgorithm. The sorting problem was solved by the selection sort.

9.5. EXERCISES

9.01. Modify the search subalgorithm to allow specification of an additional precondition: the position within the array at which the search is to start. This allows the search to begin at a position other than 1. What error conditions must be checked for? How should you resolve these errors?

9.02. Using the modified search subalgorithm from exercise 9.01, write a subalgorithm that counts the number of times a particular value occurs in a list. Assume you are given an array that already contains the values in the list, the array's placekeeper, and the particular value. Your subalgorithm should return only one postcondition: the number of times the value was found in the array.

Is it better to use the modified search subalgorithm or write a new subalgorithm to solve this problem? Give your reasons.

9.03. As an extension of exercise 8.08, modify the algorithm to read transaction information from the terminal in a questions-and-answer sequence. You will read the initial account information from a data file, but the data file will be slightly changed.

Each account line is followed by another line, giving the customer's secret authorization code and name. The format is:

CCCC LLLLLLLLLL FFFFFFFFFF

where CCCC is the secret authorization code

LLLLLLLLLL is the last name (≤10 chars)
FFFFFFFFFF is the first name (≤ 10 chars)

Sample Data

178996	1.00	1.25
1732 Washington George		
186165	16.00	5.01
1809 Lincoln Abraham		
000000	0.00	0.00

You will probably find it easier if the interactive portion of your algorithm is confined to a subalgorithm. You simply replace 1 read in your algorithm with a call to the subalgorithm.

Sample Input/Output

Welcome to Jayhawk National Bank.
Please enter your account number.
=186165
Hello, Abraham Lincoln
Please enter your secret authorization code.
=1809
Would you like to make a Credit or Debit?
=Yes
Sorry, "Y" is not a legal transaction code.
Would you like to make a Credit or Debit?
=C
Do you want your Checking or Savings?
=S
Please enter the transaction amount.
=25.00
Credit savings with $ 25.00
Your new balance is $ 41.00

9.04. This exercise deals with arrays and "searching." Write an algorithm to implement a seat-reservation algorithm for an auditorium that has 30 rows of seats and 10 seats in each row. The rows are numbered 1 to 30 (front to back) and the seats 1 to 10 (left to right, facing the stage). For selling tickets, the auditorium has been divided into 4 sections, with seats in each section differing in price. The division is done on a row basis, as follows:

Section Name	*Rows*	*Price*
FRONT	1-3	$4.00
PREMIUM	4-15	$6.00
MIDDLE	16-25	$5.00
BACK	26-30	$3.50

The data your algorithm should read consists of requests for seats by section (e.g., FRONT 3 requests 3 seats in the section named FRONT). You do not have to get all the seats together to get full credit for this exercise.

The last data line has a section name of END. After all requests have been filled, print a seating chart of the auditorium with a period (".") for an empty seat and an "X" for a filled seat (because the seating chart has 30 rows of 10 seats each, you print 30 lines with 10 periods and X's).

Your algorithm should print a message that indicates where each reservation was (or was not) placed (and if not, for what reason).

Possible extension: Fill every request with adjacent seats, if possible. If this is not possible, fill the request anyway—but print a message to this effect.

9.05. Write a primitive version of the algorithm that produced this book. Looking at the right-hand margin, you see that each line is the same length, and this is achieved by adding blank spaces throughout until each line is the same length. This process is called **justification**.

The first line of the data file contains the desired output line length (less than or equal to 60); the remaining lines contain text that should be justified. Each data line contains a series of words, separated by blanks. None is longer than the desired output line length. Justify each line and output the justified text.

There are two exceptions. If a line is less than half the desired output line length, it is assumed to be the end of a paragraph and therefore should not be justified. Also, if a line consists of only 1 word (regardless of length), it obviously cannot be justified.

In addition, if a line starts with 1 or more spaces (i.e., is indented), you should preserve the indentation in the output line.

Sample Input

```
25
   This is some data
to show what the output
from this exercise
might look like.
```

Sample Output

```
   This  is  some  data
to show what the output
from this exercise might
look like.
```

9.06. Write an algorithm for making airline reservations. Assume you are working for a travel agency that sells tickets only for TWA and United Airlines.

The first 10 lines in the data file contain information about 10 flights departing from Kansas City International Airport tomorrow. Each line has a flight number (6 characters), the number of seats still available on that flight, departure time (in 24-hour time), and destination (15 characters).

Sample Input

```
TWA280 12 17.30 CHICAGO
UA566C  4  8.15 DENVER
. . .
```

The customer list begins on line 11. Each line has a customer's name (15 characters), departure time, and destination (15 characters). The data file ends at the end of the file.

Sample Input

```
REX JENKINS      8.00 DENVER
DAVID CHERITON  17.10 CHICAGO
```

Give a customer a reservation if a seat is available on a flight that goes to the desired destination and if departure time is not more than an hour after the requested time. Then output all the names of passengers who are given reservations and their flight numbers.

Sample Output

```
          Passenger List
NAME                      FLIGHT
--------------------------------
REX JENKINS               UA566C
DAVID CHERITON            TWA280
```

Also output a list of customers who are/were unable to get a reservation.

9.07. You are head programmer for a major Midwestern wholesale distributor that has 6 warehouses and sells 5 different items. Each warehouse may stock any or all of the 5 items; in addition, your company buys and sells these items with great regularity. Company sales take the form of transaction lines that contain a transaction code ("P" for a purchase, "S" for a sale), an item number (1, 2, 3, 4, or 5), and the quantity bought or sold. Process these transaction lines as described below. The last data line will have a transaction code of "X".

1. The first 6 data lines contain the initial status of all the warehouses, and each of these 6 lines has 5 numbers. The first line of the file is the status of warehouse 1 and the first number is the quantity of item 1 on hand; the second number is the quantity of item 2; etc. Write the initial status of the warehouses.

2. All data lines after the initial 6 are transaction lines, and the following subalgorithms process each transaction:

   Sale—a subalgorithm that subtracts the quantity sold from the warehouse that has the largest supply of the item.

   Purchase—a subalgorithm that adds the quantity purchased to the warehouse that has the lowest supply of the purchased item.

   Your output should include the information from the transaction line, as well as designate the warehouse. If there is an error on a line, print an error message; but do not process the line.

3. After all transaction lines are processed, print the final status of all the warehouses.

9.08. Develop an algorithm that reads in two numbers, N and M, where N is the number of rows and M is the number of columns in a matrix; that reads in two N × M matrices; that adds the two matrices; and that prints out the two addends and their sum (3 matrices). Repeat this until you run out of data.

Addition of matrices: The sum of two N × M matrices, A and B, is an N × M matrix, C, where each element of C is given by the formula:

$$C_{ij} = A_{ij} + B_{ij}$$

with $1 \leq i \leq N$ and $1 \leq j \leq M$.

Sample Input

```
2 4
1 0 3 4
5 8 4 6
9 1 0 2
6 8 4 3
```

Sample Output

```
1  0  3  4
5  8  4  6

     +

9  1  0  2
6  8  4  3

     =

10   1  3  6
11  16  8  9
```

9.09. Develop an algorithm for the game tic-tac-toe, to be played by two players. The game starts with an empty board (3 by 3 array). The players take turns selecting the square they wish to occupy. After each move, the status of the board should be printed and the next player prompted for input. As each square is filled, your algorithm should check for a winner: 3 squares vertically, horizontally, or diagonally, occupied by one of the players. After each game, print a message (who won), and ask if they want to play another game. The players alternate as to who moves first in each game.

Turn in a copy of your algorithm and the output of at least two games.

Sample Input/Output

```
Your move, X.
=1 3
- - X
- - -
- - -

Your move, O.
=3 3
- - X
- - -
- - O

Your move, X.
=1 1
```

```
X - X
- - -
- - O
  .
  .
  .
X O X
X O -
X - O

Congratulations player X

Do you want to play again?
=Yes

Your move, O.
  .
  .
  .
```

9.10. This problem is essentially the same as exercise 9.09, except that your algorithm is for one of the players. It can use any strategy you choose, with one exception: if a winning move is available to your algorithm, it must make that move.

Sample Input/Output

```
You are player O.
The computer is player X.

I move 2 2
- - -
- X -
- - -
What is your move?
=2 3
- - -
- X O
- - -

I move 1 3
```

```
.
.
.
X O O
- X O
X - X

I win!!!

Do you want to lose another game?
=No
```

9.11. Producing a calendar for any year, your algorithm should read the year from the terminal. Your instructor should provide a subalgorithm to calculate the first day of the year.

Sample Output

```
JANUARY                                        1981
---------------------------------------------------
        S    M    T    W    T    F    S
---------------------------------------------------

                             1    2    3
        4    5    6    7    8    9   10
       11   12   13   14   15   16   17
       18   19   20   21   22   23   24
       25   26   27   28   29   30   31
---------------------------------------------------
```

Possible extension: Produce a fancy calendar, something like

J A N U A R Y . . .

SUNDAY	MONDAY	TUESDAY	. . .
			. . .
4	5	6	. . .

If you want, emphasize holidays and the birthdays of your friends and relatives, or produce a calendar with special marks of your own.

To show that your algorithm works, produce a calendar for the current year.

9.12. Members of Densa (a high-IQ society) have invented a new game, a cross between blackjack and poker. If all the players get 21 points or go bust, the best poker hand wins (instead of the dealer). However, they're having problems winning at this game (except by blind luck). To help them out, you write an algorithm to calculate probabilities.

The players type into your algorithm each card that is dealt them; also, they can ask you to calculate a probability at any time. Your algorithm should be able to calculate the probability of getting one card of a set of cards. Players should be able to ask for probabilities for a specific card (Ace of Spades), a card of a certain color or suit (a red, a club), a specific card of a certain color (a black jack), or, since the idea is *not* to break 21, any card under an amount (4 or less).

$$\text{Probability of event} = \frac{\text{\# of times event occurs}}{\text{\# of possible events}} = \frac{\text{\# of matching cards}}{\text{\# of cards remaining}}$$

Your input should be formatted into three fields: info, suit, and rank. The first tells if exactly that rank is wished, or that rank and lesser ranks. The next specifies the suit (or group of suits) the card should belong to. The last field (a 2-digit integer) specifies a card's rank.

```
Info = {E: Exactly,
        L: That card or less}
Suit = {A: Any suit,
        B: Black (Club or Spade),
        C: Club,
        D: Diamond,
        H: Heart,
        R: Red (Heart or Diamond),
        S: Spade}
Rank = {0: any rank, 1 to 13: corresponding ranks}
```

This is illustrated by the following examples:

Input = *Probability of Drawing . . .*

EB11	A black jack
LA 4	A four or less
LH 9	A heart, nine or less
ER00	A red card
LS 0	A spade
EC07	The seven of clubs

Your algorithm must be interactive and should allow the players either to calculate a probability or type in a dealt card until they want to stop playing. Also, they should be able to tell your algorithm that they have shuffled the deck.

9.13. Modify the algorithm you wrote for exercise 8.04 to allow an instructor to verify or correct student data at the terminal. You read the initial student information from a file, then allow the user at the terminal to add a student, print out information for a student, and correct the information for a student. When all corrections have been made, compute the grades (as you did before). Allow all the capabilities in the sample input/output sequence below. The sample data file is the same as before.

Sample Input/Output

```
There are 3 students.

Add, Print, Change, or Quit?
=Add
Enter last name and first name:
=TSS Joe
Enter ID number, class, and school:
=176328 2 Z
Enter the three test scores:
=90 80 70

Add, Print, Change, or Quit?
=Print
Enter ID number:
=111111
Name = Ofarc Joan
ID = 111111, Class = 4, School = A
1st test = 90, 2nd test = 81, Final = 98
```

```
Add, Print, Change, or Quit?
=D
Sorry, "D" is not a legal response.
Add, Print, Change, or Quit?
=C
Enter ID number:
=11111
Sorry, ID number not found: 11111
Enter ID number:
=111111
Enter item to change (Name, 1st test, 2nd test, Final):
=1
Old score was 90, enter new score:
=91

Add, Print, Change, or Quit?
```

9.14. Write an algorithm to play the game of Mastermind. (On the computer we play with numbers instead of colors.) One person types in a 4-digit number as a "secret number" and another person tries to guess it. The number consists of digits between 1 and 9 (and no repeated digits).

Each time a player guesses, print a response that indicates how close to the number the guess is. Print an "X" for each digit that's in the right place, and an "O" for each digit that's in the wrong place (but appears in the "secret number" in another place). Do not line up the X's and O's under the digits that are correct or present (i.e., print all the X's, then all the O's). The idea is to tell the player *how many*, not *which*, digits are correct or present.

The game ends after 10 unsuccessful guesses, or earlier if the number is guessed.

Sample Input/Output

```
Enter the secret number and then clear the screen.
=1492

Make a guess.
=1234
XOO

Make a guess.
=9876
O
```

```
Make a guess.
=1298
XXO

Make a guess.
=1492
That is the secret number!
You got it in just 4 guesses!
```

Possible extension: Have the computer pick the secret number. Your instructor should provide a subalgorithm to generate at random a 4-digit number between 1111 and 9999. Check to make sure it has no zeros or duplicated digits.

9.15. In an algorithm that ranks poker hands, each line on the data file contains information for a single hand. The format is

```
SRR SRR SRR SRR SRR
```

where S is a suit ("C" for clubs, "D" for diamonds, "H" for hearts, "S" for spades), and RR is a rank (1-13). Aces (1) rank above kings (13), and hands in poker are ranked as follows:

1. One pair
2. Three of a kind
3. Two pair
4. Full house (three of a kind and a pair)
5. Straight (cards in sequence, like 7-6-5-4-3)
6. Flush (all cards of the same suit)
7. Straight flush
8. Four of a kind

Write an algorithm that classifies each hand. Possible extension: Print out which hand in the data is highest. Ties are broken by comparing ranks; that is, the cards that make (say) a pair are compared for superior ranking, and then (if need be) the other cards are compared (also by rank). If ranks are identical, suits are compared by rank (in descending order: spades, hearts, diamonds and clubs).

9.16. Write an algorithm that checks the words in a text against a dictionary— a list of correctly spelled words. (Both the dictionary and the text have been entered 1 word per line to make it easier.) The entire dictionary should be read in before you check the text, because the dictionary is not organized in any particular order. Every word in the text that is not in the dictionary should be printed, along with the word in the dictionary that is the "closest match" (as defined below).

Nonidentical words are compared character by character, and the number of different characters is counted. The word with the least number of differences is the closest match, as long as this difference is less than 40% of the length of the word. For example:

Text:	TEH	KNIVES	NIGHT	NITE	NITE	CHAIRS
Dict:	THE	KNIFE	KNIGHT	NIGHT	LITE	CHAIR
Count:	2	2	6	3	1	1
	66%	33%	120%	75%	25%	16%
		Yes			Yes	Yes

Sample Text

THE
GRANDE
APPLE
APPPLE
BANANNA
FRUTE

Sample Dictionary

APPLE
ORANGE
BANANA
FRUIT

Sample Output

Misspelled words	Possible correction
THE	(NONE)
GRANDE	ORANGE
APPPLE	(NONE)
BANANNA	BANANA
FRUTE	FRUIT

5 of 6 words were misspelled.

9.17. As a programmer for the Smoky Publishing House, you have been asked to write an algorithm that will perform Smoky's Readability Test. SRT is a test that analyzes the reading level of a passage of text by counting the lengths of the words. Print out the total number of words in each length category and each category's percentage of words in the entire passage.

Assume that there are no hyphenated words.

Sample Input

As one grows in the art of computer programming, one constructs programs in a sequence of refinement steps. At each step the programmer breaks his/her task into a number of subtasks, thereby defining a number of partial programs. Although it is possible to camouflage this structure, this is undesirable. The concept of the procedure allows the display of the subtasks as explicit subprograms.

Sample Output

As one grows in the art of computer programming, one constructs programs in a sequence of refinement steps. At each step the programmer breaks his/her task into a number of subtasks, thereby defining a number of partial programs. Although it is possible to camouflage this structure, this is undesirable. The concept of the procedure allows the display of the subtasks as explicit subprograms.

Analysis of the above text

Word Length	Number of Words	% of Total
1-3	29	45.313 %
4-6	12	18.750 %
7-9	16	25.000 %
10-12	7	10.938 %
over 12	0	0. %

10
Efficiency and Analysis of Algorithms

In Chapter 5 we introduced the notion of analyzing an algorithm according to its input length, N. Accordingly, the machine must have the necessary accuracy, speed, and memory capacity to execute a program correctly in a reasonable amount of time. On the other hand, we have done very little about the efficiency and analysis of algorithms, primarily because there's no need to worry about efficiency unless a program is correct. Thus far, the emphasis has been on "correct." *Now* we're concerned with analyzing algorithms.

Efficiency is a relative term—one algorithm is more efficient than another if its cost or size of execution is less than another's. In the same way that the size of execution is defined by its input length, the size of an array is defined by its number of elements. There are two basic factors related to cost: use of storage and use of computer time. A third factor, which indirectly affects cost, is simplicity and generality.

In general, the cost of running a program is directly related to the amount of storage (memory) and computer time (CPU time) used. The more storage, the higher the cost; the more computer time, the higher the cost. Our objective for an efficient algorithm is minimum storage and minimum execution time.

This is easy to say but difficult to accomplish. In many problems, only one of these objectives is possible.

10.1. USE OF STORAGE

Computer memory consists of addressable locations or cells. When we construct an algorithm to run on a computer, we want to use a few locations and make as few memory references as possible. To utilize minimum storage, we must select and arrange data structures accordingly. In Chapter 8, we introduced the concept of aggregate data types and in particular, an array; and we know that in designing an algorithm it may be necessary to use two or more arrays of the same size. If each array could share a memory location, this would reduce the storage required—if the program's two arrays are not needed simultaneously.

Many programming languages handle this automatically by allocating and freeing memory as a program is executed. High-level languages that have this capability are called **block structured** languages. If a language has this capability, there are again tradeoffs that must be considered. It typically takes more computer time to accomplish block structure capability. Hence, we may have decreased the amount of storage, but we have increased the amount of computer time.

10.2. USE OF COMPUTER TIME

Execution efficiency is extremely complex, and many of its principles go beyond the scope of this text. Some of this is due to the fact that more research needs to be done in the area. One of the major concerns in execution time is the time it takes for the machine (computer) to access a memory location (**memory access time**). From Chapters 3 and 5, we saw that a program is stored and instructions are referenced as execution is carried out. Therefore many high-level languages have "optimizer" versions of a compiler that reduce the number of instructions needed to execute.

Many inefficiencies creep into implementation of an algorithm because of redundant computation or unnecessary storage. In fact, redundant computation is most noticeable when it is embedded within a loop. For example, a common mistake is recalculation of an expression that remains constant through the entire loop. This is especially true in referencing array elements, because at least two memory references are needed. Many compilers do trivial optimization, such as moving constants out of a loop. In general, let the compiler do the simple optimization. (This is discussed in more detail in Appendix A.)

Another simple way to speed up an algorithm is by using the least number of loops. (The major drawback is that this makes the eventual program harder to read and debug.) In general, it is best to make one loop do only one job. When you select an efficient solution, it is better to improve the algorithm and not resort to programming "tricks" (which tend to obscure what is really being done).

Computer time can be a criterion in selecting the best algorithm. When an algorithm is seen to be overly time-consuming, use a comparison table to select the best method or algorithm. For example, consider Figure 10–1, which compares two methods of solving the same problem. If the number of data values is small (say <25), method 1 is faster. If the number of data values is approximately 50, either method is good. Clearly, for a large number of data values, method 2 is better.

	Number of data values		
	10	100	1000
Method 1	3 sec.	48 sec.	456 sec.
Method 2	50 sec.	50 sec.	50 sec.

FIGURE 10–1 Comparison of Two Algorithms

Careful analysis of algorithms can have a strong effect on the efficiency of a program.

10.3. ANALYSIS OF ALGORITHMS

As we have seen, there are many ways to solve a problem or program. In Appendix A, we will measure the "quality" of a program (program stylistics). It is very easy for us to see that we are interested in solutions that are economical in their use of computing and human resources. Economy, of course, is all well and good, but we must also have some quantitative measures to evaluate the "goodness" of an algorithm: to predict its performance and, especially, compare the relative performance of two or more algorithms.

To provide a quantitative measure of an algorithm's performance, it is necessary to set up a computational model that describes algorithmic behavior under specific input conditions. As described in Chapter 5, we do this in terms of the size (usually N) of the problem. In the matrix example (Chapter 5), we saw that if x is the time required to process a list of size $N = 1$, N^2x is the time required to process a list of size N. Hence, the time required is *not* linear, 10x, but quadratic: $10^2x = 100x$. In general, the important question is how the cost of solving the problem varies as N increases.

As we evaluate the computational cost of many simple algorithms, we see that the cost ranges from simple linear dependence on N to the high end of the scale, where dependence on N is exponential (2^N). Figure 10–2 illustrates the comparative costs for a range of values of N.

It is obvious that we can solve problems for only small N when the algorithm exhibits exponential behavior. An example is the traveling salesman problem (described in Chapter 5). If a step is 1 microsecond long, then when N is large (e.g., 10^4), the difference between $N\log_2 N$ and N^2 would be significant.

N	$\log_2 N$	$N\log_2 N$	N^2	2^N
2	1	2	4	4
10	3.322	33.22	10^2	1024
10^2	6.644	664.4	10^4	$>10^{25}$
10^4	13.287	132,877	10^8	$>10^{2500}$

FIGURE 10–2 Comparative Costs for a Range of Values

The next step is to understand an algorithm well enough to determine what represents N. In the matrix transpose problem (Chapter 5), this was fairly easy since the dominant mechanism was the number of times an interchange of elements must be made. In the next section, we will look at a few sorting and searching algorithms and see that comparison, exchanges, and moves characterize the dominant mechanisms.

The notation by which we evaluate algorithms is the **O-notation** ("Big-O"). In this notation, an algorithm in which the dominant mechanism is executed KN^2 times for a constant K and problem size N is said to have *order* N^2, which is written $O(N^2)$. Formally, a function g(N) is O(h(N)), provided there is a constant K for which the relationship

$$g(N) \leq Kh(N)$$

holds for all values of N that are finite and positive.

With this relationship we have a means of determining the asymptotic complexity of an algorithm. We can write the above inequality in the form

$$\lim_{N \to \infty} \frac{g(N)}{h(N)} = K, \text{ where } K \neq 0$$

For example, if we determine that $g(N) = 10N^2 + 5N + 100$ for an algorithm and that

$$\lim_{N \to \infty} \frac{10N^2 + 5N + 100}{N^2} = 10$$

it follows that, since $h(N) = N^2$ and $K = 10$, this algorithm has asymptotic complexity of $O(N^2)$.

In the matrix transpose algorithm of Chapter 5, $g(N) = N(N-1)/2$. Since

$$\lim_{N \to \infty} \frac{N(N-1)/2}{N^2} = 1/2$$

it follows that $h(N) = N^2$ and the asymptotic complexity of the algorithm is $O(N^2)$.

In the simple averaging algorithm of Chapter 5, $g(N) = N$. Since

$$\lim_{N \to \infty} \frac{N}{N} = 1$$

it follows that $g(N) = N$ and the asymptotic complexity of the algorithm is $O(N)$.

In a more complex example, let $g(N) = 2^N + N^2$. If we let $h(N) = 2^N$, the value of K is 1, because 2^N increases faster than N^2. Hence the

$$\lim_{N \to \infty} \frac{2^N + N^2}{2^N} = 1$$

Hence an algorithm with $g(N) = 2^N + N^2$ has asymptotic complexity of $O(2^N)$. Many authors refer to this asymptotic behavior as **worst-case** complexity. Another concept that is mentioned (but will not be discussed in detail in this book) is *expected* or *average* complexity. Expected complexity yields a measure of the algorithm, averaged over all possible problems of size N. If both the expected and worst-case complexities are available in the evaluation of algorithms, we always use the expected complexity. The problem is that expected complexity usually involves very complicated combinatorial analyses and is usually not available.

10.4. THE SORTING PROBLEM

Sorting algorithms arrange (sort) items in sets according to a predetermined relation. The two most common types of data are *numerical* and *character* information. The ordering relation for numerical data involves arranging items in sequence from largest to smallest (descending) or from smallest to largest (ascending). For example, the items in the set {1, 5, 9, 13, 18, 22} are arranged in ascending order. String information is usually arranged in standard lexicographical or dictionary order. For example, the items in the set {b, bill, billed, g, gore, greg} are in lexicographical order.

Because of the importance of sorting, many algorithms have been studied extensively, and usually fall into two classes. The first is simple algorithms, which we characterize by the fact that they require N^2 *comparisons* to sort N items. Their worst-case behavior is $O(N^2)$. The second class of algorithms requires $N\log_2 N$ comparisons to sort N items. These algorithms come close to the optimal set because their worst-case behavior is $O(N\log_2 N)$. These algorithms differ in how they utilize storage and in their expected behavior.

In the ratio $N^2/(N\log_2 N)$, we see that if N is small ($N = 10$), the ratio is 3.01 and not very significant. But if $N = 10{,}000$, the ratio is 752.58, which is very significant.

10.4.1. The Bubble Sort Algorithm

Problem: Sort an array of elements $a_1, a_2, \ldots, a_N$ into an ascending single-ordered array.

```
Start Algorithm

   read N and array a1, a2, . . . , aN
   print N and the array a
   set switch to true
   while switch = true do
```

```
        set switch to false
        set J to 1
        while J ≤ N−1 do
          if a_J > a_J+1
            then
              interchange a_J and a_J+1
              set switch to true
          end if
          set J to J+1
        end while
      end while
      write, the sorted array a_1, a_2, . . . , a_N

    End Algorithm
```

ALGORITHM EXAMPLE 10–1

Algorithm 10–1 (for bubble sort) consists of passing through an array, comparing each successive pair of elements, interchanging pairs that are out of order, and repeating this process until the array is in order. A unique feature in the algorithm is the ability to stop when the array is sorted, by using the variable switch. This feature makes the algorithm slightly more efficient by ascertaining whether exchanges have been made in the current pass. Thus we determine whether the array has been sorted. If there have been no exchanges, it follows that all elements in the unsorted part of the array must already be in ascending order.

This version of the bubble sort is inefficient, because *every* element is checked in the inner while-do. A more efficient algorithm can be constructed by observing the $N-J+1$ element (exercise 10.01). The time required to sort an array of N elements, using a bubble sort, is still proportional to N^2. This makes the bubble sort efficient only for small values of N. It is recommended that the reader sort a small array of numbers, using the algorithm and pencil and paper. Many refinements can be made to the bubble-sort algorithm, but it still remains an $O(N^2)$ algorithm.

If we ignore some of the less important details of the bubble sort, we find that the time to execute it is

$$\frac{N(N-1)}{2} t_L$$

where t_L is the time required to make the interchange in the inner loop. This measure is sufficient since we are primarily interested in comparing algorithms, not in finding an exact time for a specific implementation.

$$\frac{N(N-1)}{2} t_L = \left(\frac{N^2}{2} - \frac{N}{2}\right) t_L$$

or approximately

$$\frac{N^2}{2}\, t_L$$

From this we write

$$g(N) = \frac{N^2}{2} \text{ or } O(g(N)) = N^2$$

Additional algorithms perform the sort at the lower bound of the problem for worst and expected cases of $O(N\log_2 N)$ for N items. Examples include *heap sort* and *merge sort* (Knuth; Aho, Hopcroft, and Ullman; Horowitz and Sahni). *Quicksort* has the best expected behavior, but a worst-case behavior $O(N^2)$. Because of its excellent expected behavior, the reader is encouraged to analyze the algorithm in more detail.

10.5. THE SEARCHING PROBLEM

Searching algorithms search an ordered file or set to determine whether an element is or is not present—a dictionary, say, or a telephone directory. For example, one way to find a name in a telephone directory is to start at page 1 and continue page by page until we are successful *or* have examined all the pages. If we have exhausted all the names, there is no such name in the directory. If the directory is very small, this method may be practical; if the directory is large (like New York City's), the method is not practical. The algorithm for accomplishing the above is called *sequential search.*

10.5.1. Sequential Search Algorithm

Problem: Search an ordered (ascending and numerical) set, S, to determine whether element X is present. Denote the set by $(K_1, K_2, \ldots, K_n)$.

```
Start Algorithm

  read the set S=(K1,K2, . . . ,Kn)
  read X
  set i to n
  while (Ki ≠ X) and (i ≥1) do
    set i to i-1
  end while

End Algorithm
```

ALGORITHM EXAMPLE 10-2

Algorithm 10–2 passes through a set ($K_1, K_2, \ldots, K_n$). If a match is found, $K_i = X$. If a match is *not* found, i is set to zero, and this signifies that a match was not found.

It should be obvious from the above that the worst-case behavior is O(N), since it is possible to search the entire list. If, after all uses of this algorithm, all key elements K are searched for an equal number of times, then, on average, N/2 comparisons will be made before an element is found. It follows, then, that g(N) for average behavior should be approximately half the g(N) for the worst-case behavior. Hence the average behavior is still just O(N).

Other algorithms perform searches. Examples include the *binary* search, the *Fibonacci* search, and the *interpolation* search. Because the binary search has one of the *best* worst-case behaviors, namely $O(\log_2 N)$, let's discuss its algorithm in more detail.

Again, we are searching an ordered ascending numerical set, S, to determine whether an element, K, is present. It is called a *binary* search because we always compute the index of the middle of a set and determine whether our key is greater or less than the element of the set with that index. If the key is equal to the set element with that index, the search has been successfully completed.

As an example, consider the set

$$S = (0,2,10,13,17,20,21,29,32,34,35,39,47,53,60)$$

The set size is n = 15. We compute by averaging the index of the smallest and largest element and truncating the answer. In this example, the average would be (1 + 15)/2, which is 8. This is usually denoted by trunc((1 + 15)/2) = 8. The element with the index 8 is 29. If the element we are trying to find has a value greater than 29, we search the set of elements greater than 29, namely, (32,34,35,39,47,53,60). If the element we are trying to find has a value less than 29, we search the set (0,2,10,13,17,20,21). Because this splitting process can be described by a binary decision tree, it is called a binary search.

If the element we are seeking were in the set (0,2,10,13,17,20,21), the next index would be trunc((1 + 7)/2) = 4 in S. If the element we are seeking were in the set (32,34,35,39,47,53,60), the next index from S would be trunc(((8 + 1) + 15)/2) = 12. The reason for the (8 + 1) is that the element 32 in set S has the index 8 + 1 = 9.

The binary tree for indexes of this process is shown in Figure 10–3. Thus we make only three comparisons to see if an element is in the list.

10.5.2 Binary Search Algorithm

Problem: Search an ordered (ascending and numerical) set S to determine whether an element X is in the set. Denote the set by ($K_1, K_2, \ldots, K_n$).

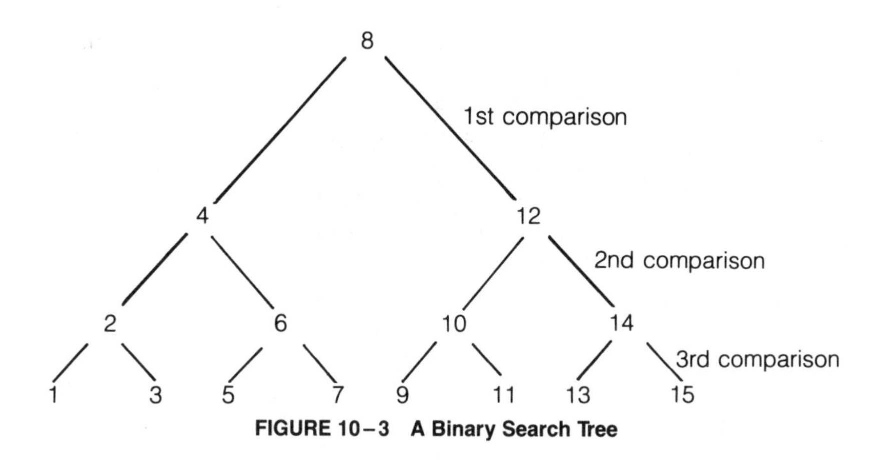

FIGURE 10–3 A Binary Search Tree

```
Subalgorithm to search an array,
                given an element, X, the array, K, its placekeeper, n, and
                returning the position, i, of the match,
                    or zero if no match is found.

  set i to 0
  set j to 1
  set u to n
    while j ≤ u do
      set avg to the truncated value of (j+u)/2
      if X > K_avg
        then
          set j to avg+1
        else
          if X = K_avg
            then
              set i to avg
              set j to u + 1
            else
              set u to avg-1
          end if
      end if
      end while

End Subalgorithm
```

ALGORITHM EXAMPLE 10–3

As in the sequential search, $K_i = X$, if a match is found. The index, i, points to the element found. If a match is not found, i is set to zero. In this algorithm, the worst-case behavior is $O(\log_2 N)$.

Consider how we determined that the algorithm has worst-case behavior logN. After one comparison, the set or file to be searched has length N/2; after the second, it has length N/4; and so forth. If the set length, N, is initially 2^m, then, after m comparisons, the set to be searched has length $N/2^m$. Eventually, the set to be searched will have only one element (the right one).

We might have found a match much earlier, but we are interested only in the worst case. Hence, for the worst case, we have a value of m such that

$$\frac{N}{2^m} = 1$$

or

$$m = \log_2 N$$

Thus the worst-case behavior for this algorithm is $O(\log_2 N)$.

If we look at the ratio of $N/\log_2 N$, we see that if N is small (e.g., N = 10), the ratio is 3.01. If N is large (e.g., N = 10,000), the ratio is 752.58, which is very significant. Clearly, binary search is better than sequential search.

10.6. CONCLUSIONS

From the above we have seen that the efficiency of an algorithm is closely related to design and implementation considerations, as well as the analysis of an algorithm. Each algorithm utilizes many computer resources to complete its work. The most relevant include the time of the central processing unit and the amount of main memory. These are the two important components of algorithm efficiency.

Like most aspects of algorithm design, there is no easy recipe for designing an efficient algorithm. Despite the drawbacks, helpful generalities for the analysis of algorithms are implicit in worst-case behavior and the average behavior of an algorithm. These characteristics help us draw rational conclusions about the efficiency of an algorithm.

10.7. EXERCISES

10.01. In the bubble sort algorithm, it can be observed that, after pass J, the Jth largest value is in element [N − J + 1]. Hence it is only necessary to examine elements with indexes less than N − J + 1 during the next pass. Modify the algorithm to take advantage of this.

10.02. If $g(N) = 10$ for an algorithm, show that it has order unity, written as $O(1)$.

10.03. If $g(N) = 3^N + N^3$ for an algorithm, show that it has order 3^w, written as $O(3^w)$.

10.04. If $g(N) = 2 + 1/N$ for an algorithm, show that it has order unity, written as $O(1)$.

10.05. Find the order of $g(N)$ if $g(N) = N! + N^2$.

10.06. Plot the order for both the binary search algorithm and the sequential search algorithm on graph paper. What is your conclusion?

10.07. Create a table similar to Figure 10–1, where method 1 is the bubble sort and method 2 is another sort that you must look up in the literature. You must implement each algorithm in a specific programming language.

10.08. Repeat the last exercise, except use the binary search algorithm as method 1, and the sequential search algorithm as method 2.

10.09. Create a table similar to Figure 10–1, where the method is an algorithm you have implemented during this course. You do not need to have a method 2.

10.10. Modify the bubble sort algorithm to sort the array A in descending order (largest number first).

10.11. When does the worst case occur in the bubble sort algorithm?

Appendix A
Program Stylistics

A.1. CLARITY

A.1.1. Writing Clearly

Clarity and simplicity in writing algorithms make it that much simpler to test a program for correctness, as well as to modify code. This is especially important because most programs that are used in business and industry must be changed frequently in response to new laws, new management policies, new products, etc.

A.1.2. Comments

The purpose of inserting comments in a program is to help those who need to modify an unfamiliar program (and *you* may be one of these people). Most programmers, nevertheless, have had the unsettling experience of reading a program that they themselves wrote and being unable to discern *what* the program does.

A properly commented program should contain (at minimum) a preface and an explanation of the use and purpose of each variable (as well as any statement whose intent is not obvious). In addition, it is often helpful if comments help match begin-end pairs and clearly delineate blocks.

The following are the algorithm and a Pascal implementation of a well-commented Celsius-to-Fahrenheit conversion. The algorithm is straightforward:

```
Start Algorithm
  print headings
  for CelsiusTemperature ← −20 to 80 do
    compute the Fahrenheit equivalent
    print both temperatures
  end for
End Algorithm
```

ALGORITHM EXAMPLE A–1

The Pascal program that corresponds to Algorithm A–1 is

```
program Conversion (output);
{ Program to convert temperatures from −20 degrees to 80 degrees
  Celsius to equivalent temperatures in degrees Fahrenheit. }
const
  Low = −20;
  High = 80;
var
  Fahrenheit : real;
  Celsius : integer;
begin
  { print headings }
  writeln (' CELSIUS FAHRENHEIT');
  writeln;
  { compute and print the temperatures }
  for Celsius := Low to High do
    begin
      Fahrenheit := (9/5)*Celsius + 32;
      writeln (Celsius :6, Fahrenheit :16:1)
    end {for}
end.
```

ALGORITHM EXAMPLE A–2

A.1.3. Mnemonic Variables

The name of a **mnemonic variable** reflects the meaning of the information it holds. For example, "totalwords" might be an integer that represents the total number of words in a paragraph, or "top" might be a pointer to the top of a list. The meaning of a program is much more easily discerned if good mnemonic variables are used throughout.

It is advisable not to use variable names that are similar to each other. For example, don't use COUNTER2 and COUNTERZ in the same program.

A.1.4. Indentation

In as much as we allow a free-format language, we can understand an algorithm whether or not indentation is used. However, an algorithm that uses a good indentation scheme is much easier to read and understand. The following examples of algorithm segments that behave in the same fashion will highlight this point.

```
Start Algorithm segment

  set done to true
  for i ← first to last − 1 do
    set big to list_i
    set place to i
    for j ← i + 1 to last do
      if list_j > big
        then
            set big to list_j
            set place to j
      end if
    end for j
    set temp to list_i
    set list_i to list_place
    set list_place to temp
  end for i
End Algorithm
```

ALGORITHM EXAMPLE A–3

```
Start Algorithm
set done to true
for i ← first to last − 1 do
set big to list_i
set place to i
for j ← i + 1 to last do
if list_j > big
then
set big to list_j
set place to j
end if
end for
set temp to list_i
set list_i to list _place
set list_place to temp
end for
End Algorithm
```

ALGORITHM EXAMPLE A–4

A.1.5. Clever Code

Don't write "clever code" in an attempt to make small gains in efficiency. In the 1950s it was very important to write highly efficient code because machine time was expensive in comparison to programmer time. But in the mid-1980s the opposite is true. Programmer time is becoming much more expensive and machine time much less expensive (because of mass production and technological developments). Also, an operation that is slow on today's machine may be fast on next year's model. Concentrate on writing "*correct* code", not "clever code."

Two important corollaries:

1. Avoid complex if statements.

```
set done to length > max        if length > max
                                  then
                                    set done to true
                                  else
                                    set done to false
                                end if
```

   Although the second statement may require a few more keystrokes than the first, its intent is much more apparent than the first.

2. Use parentheses to make the order of precedence clear.

```
set answer to a + b * c and set answer to a + (b * c)
```

   In the second statement, there is no doubt about the order of operations, even if the reader has not consulted a programming manual for some time.

A.1.6. Modularity

Organize your program into major sections that are mutually exclusive (except for data). This helps organize your thoughts, as well as your program.

"The best way to achieve module independence is by"

1. Avoiding the unnecessary modification of global variables, thus making the procedure free of undesirable side effects.
2. Declaring all temporary variables local to the procedure in which they occur.
3. Avoiding changes to input parameters that were passed by reference because of memory space limitations. Whenever possible, input parameters should be passed by value so they cannot be modified.
4. Not doing anything other than what you were supposed to do as defined by the specifications of the problem. [Schneider and Bruell, p. 193]

There are always a number of ways to implement a program. Often, many different algorithms and a variety of implementations can solve a given problem, and a little thought before can save a lot of time later. For example, a comparison table can be very useful when a large number of decisions must be made—as in the following, which compares two methods of solving the same problem:

	Number of Data Values		
	10	100	1000
Method 1	3 sec.	48 sec.	456 sec.
Method 2	50 sec.	50 sec.	50 sec.

One can see from this table that if the number of data values is small (< 25), method 1 is the faster. If the number of data values is approximately 50, either method is good. And clearly, for a larger number of data values, method 2 is better.

Hence, careful analysis of algorithms can have a great effect on the efficiency of a program. These measurements are obtained by a timing routine applied to a particular method, such as

```
CALL CLOCK(Time1)
... method to be timed ...
CALL CLOCK(Time2)
```

The elapsed time, Time2 − Time1, yields the exact time the method needs. Most compilers offer such a service and the difference is usually in milliseconds.

It must be noted, however, that timing is not always sufficient. As we pointed out in Chapter 10, careful analysis of each algorithm is needed, in addition to a table of CPU times.

Let the machine do the work for you. Evaluate repetitious expressions in functions or subroutines. Do not build your own tools. Use standard library functions, such as minimum and absolute value. For example, one could write (and use) the following sequence:

```
if x < 0
  then
    set temp to -x
  else
    set temp to x
end if
```

ALGORITHM EXAMPLE A–5

Although this sequence of code works properly, it will not be obvious to the user what it does. It is easier (and much more readable) to use the following:

```
set temp to abs(x)
```

ALGORITHM EXAMPLE A–6

where abs(x) is the standard library function for computing the absolute value of x.

A.1.7. Make Your Programs Easily Modifiable

Whenever possible, parameterize your programs. This makes it easy to transport your program to another location; it also makes the program easier to read. For example, in many programming languages the logical input unit is specified as a number in the read statement. It is better to make the logical input unit a parameter where the parameter is specified before execution of the read. Instead of changing the logical input unit on every read when the program is transported to another location, all one need do is change the place where the parameter is initialized.

A.2. INPUT AND OUTPUT

A.2.1. Echoing Input to Output

Always echo print your input: use headings to identify what your program thinks is input. Make your output format as pleasing to read as possible. The better it looks, the easier it is to use.

Test input for plausibility and validity: make your program handle any erroneous data that could bc input. Give good diagnostics (whenever feasible) and make recovery easy (when possible). Make sure the input is within the limitations of the program before you proceed.

For example, consider this sequence of code, which yields good diagnostics when the input is outside the limitations of the program.

```
while there is data left do
  read, TestScore
  if (TestScore < 0) or (TestScore > 100)
    then
      write, 'The test score', TestScore,
            'is not within range'
    else
      [process the test score]
  end if
end while
```

ALGORITHM EXAMPLE A–7

Clearly, this program has the advantage of being easy to read.

Terminate input by end-of-file or sentinel. Do not use a header value data set unless this is unavoidable (users tend to make counting errors).

Sentinels have drawbacks. If more values are added to the end of the data file, the sentinel value must be moved to be in the proper place. Therefore, the best (least error-prone) method of input termination is the end-of-file method.

A.2.2. Output

Simply printing the results of a problem is grossly insufficient; it in no way represents the solution. Printing a line of numbers is meaningless if the proper label does not indicate their meaning.

For example, the following line is meaningless:

```
67161          180          2.5          4
```

If we label the numbers "Student ID," "College Major Code," "College Grade Point Average," and "Year in School," the output *would* have meaning.

A.3. Verification

Verification—the determination that the computer-program representation of a problem is equivalent to its design statement—is necessary since there is high probability that the program will initially contain errors. In fact, verification requires approximately 30 percent of the effort to develop a program. (This includes the time spent debugging the code as it is implemented, then testing the completed code.) Thus verification is a major part of program development.

The best way to keep the number of programming errors to a minimum is to use the structured programming techniques we've discussed in this book. Construct the program in a stepwise manner, with each module fitted into the existing program after it has been carefully checked. Each module is designed so that it has no side effects. (The concepts of structured programming are the framework for this book.)

Another method of verification, called a **trace**, is performed by following the algorithm or program with trial information with known results. Because programs are divided into major modules, it is necessary to show that these modules interact properly with each other. This type of verification will reveal gross programming errors, oversights, and omissions. Furthermore, one can see how the algorithm or program works.

It is important to trace the code exactly as it is written. Do not assume that you know what it is doing. Do what the code *says*, not what you think it *should* say.

After you've coded a program, several short runs with representative input parameters can be made. Indeed, the resulting trace can be manually analyzed and can show that the program follows a logical path determined by the input parameters. However, the programmer should not accept this as the final word, since all possible input parameters cannot usually be checked. Verification of this type should be ongoing, even after the program is in production.

For example, a simple trace tool is a boolean-type variable, called *debug*, which is used as part of an if-then statement. For example,

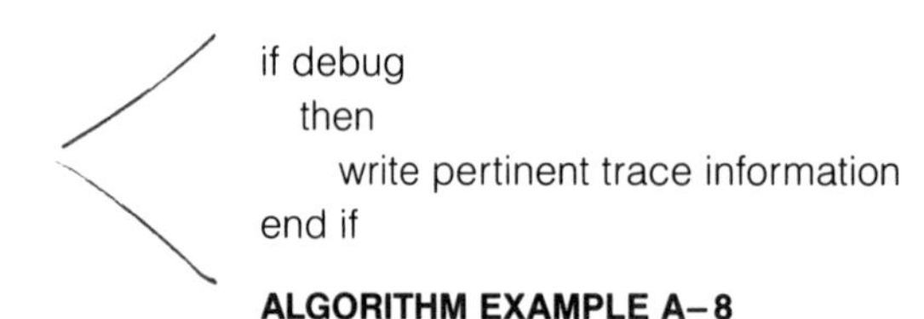

```
if debug
  then
    write pertinent trace information
end if
```

ALGORITHM EXAMPLE A–8

Hence, if debug were true, relevant trace information would be printed out; otherwise, there'd be no printing.

The only disadvantage in using this technique is that the code would appear in various parts of the compiled program and may make it more difficult to read the program.

Numerical test cases are the third tool for verification. Many programs can easily be checked by specially constructed data sets. These ensure that numerical computations are properly calculated.

This type of data set produces a predictable behavior that is monitored through the output and the trace. Any deviation between what is observed and what is expected points to a potential software problem.

A.4. EFFICIENCY

A.4.1. Program Efficiency

A program must be correct before it can be made to run faster. Many times, clarity is discarded when a program is re-evaluated for speed. Let the compiler do the simple optimizations (most compilers are excellent at this).

For example, consider the solution to the quadratic equation $ax^2 + bx + c = 0$. To calculate x, one could use the equation

set x to (−b + sqrt(b*b − 4*a*c))/(2*c)

which is very easy to program in this form. But a very common programming solution among beginning programmers is the following:

```
set p to b*b
set q to 4*a*c
set r to sqrt(p − q)
set t to 2*a
set s to −b + r
set x to s/t
```

which is less readable than the single-line solution above.

In the single-line solution, we let the compiler create the optimal way to find the solution. On many compilers, the second solution is the one the compiler will come up with. Hence the temporary variables add to the user's difficulty in reading the program.

As another example,

$$\text{set dist to dist} + (a - b)^2$$

which could have been coded as

```
set x to (a − b)^2
set dist to dist + x
```

Most compilers combine these two lines of code into one, so that it operates exactly like the one-line code. Again, it is much easier to read the one line of code.

Next, consider the sequence of code:

```
set value to x*x + sqrt(y*y + z*a − c)
if value ≤ 0
  then
    ...
end if

    ...
```

ALGORITHM EXAMPLE A–9

We could rewrite this as

```
if((x*x + sqrt(y*y + z*a - c)) ≤ 0)
  then
    ...
end if
```

If this code is embedded in a long sequence, the first (two-line) sequence is much easier to read and understand than the second. Although the second is probably more efficient, clarity should never be sacrificed.

The two major aspects of program efficiency are the space required in memory and the speed of execution. As can be seen here, program stylistics is also an important element in program efficiency.

A.4.2. Programmer Efficiency

One of the important components of programmer efficiency is the time required to design and code a program. Careful design and documentation at the beginning saves lots of debugging time.

A clear and sound design reduces the cost of maintenance, which includes expansion, modification, debugging after release, and so forth.

Good code aids the documentation. In particular, good code is the only reliable documentation of a computer program. If the code is in error, other program descriptions (pseudocode or flowcharts) will not help. Good variable names are very helpful.

A.5. DESIGN PRINCIPLES FOR PROGRAMS

Separate the task into clearly manageable modules, and make the modules as fully independent of each other as possible (minimize interfaces).

Give a functional description of each module. Do not make its interaction with other modules dependent on the implementation technique.

Carefully specify the interfaces to modules. Use standard parameter passing techniques to pass information among subprograms.

Design multipurpose routines, instead of a routine with multiple entries.

Separate system-dependent sections into a few modules. For example, centralize all I/O in a single module.

The following is a sequence of I/O-intensive code that might be placed in one module:

```
Start Algorithm Printinstructions

    Write, 'The game of checkers is played'
    Write, 'on a board with an 8×8 grid'
```

```
Write, 'on the board. Each player has'
Write, '12 pieces (of opposite, colors) at'
Write, 'the beginning of the game.'

  ...

End Algorithm Printinstructions
```

ALGORITHM EXAMPLE A–10

Again, if this is placed in one module, the program will be much easier to read.

Design your program to be compatible with other system software (use reasonable naming conventions, standard calling sequences, etc.).

A.6. SUMMARY

The following is a list of rules from Kernighan and Plauger (1974, p. 13). When in doubt, you will probably do best when you follow the rules.

Write clearly—don't be too clever.
Say what you mean, simply and directly.
Use library functions.
Avoid temporary variables.
Write clearly—don't sacrifice clarity for "efficiency."
Let the machine do the dirty work.
Replace repetitive expressions by calls to a common function.
Parenthesize to avoid ambiguity.
Choose variable names that won't be confused.
Avoid the Fortran arithmetic IF.
Avoid unnecessary branches.
Don't use conditional branches as a substitute for a logical expression.
If a logical expression is hard to understand, try transforming it.
Use data arrays to avoid repetitive control sequences.
Choose a data representation that makes the program simple.
Write first in an easy-to-understand pseudo-language; then translate into whatever language you have to use.
Use IF...ELSE IF...ELSE IF...ELSE... to implement multi-way branches.
Modularize. Use subroutines.
Use GOTOs only to implement a fundamental structure.
Avoid GOTOs completely if you can keep the program readable.
Don't patch bad code—rewrite it.
Write and test a big program in small pieces.
Use recursive procedures for recursively-defined data structures.

Test input for plausibility and validity.
Make sure input doesn't violate the limits of the program.
Terminate input by end-of-file or marker, not by count.
Identify bad input; recover if possible.
Make input easy to prepare and output self-explanatory.
Use uniform input formats.
Make input easy to proofread.
Use free-form input when possible.
Use self-identifying input. Allow defaults. Echo both on output.
Make sure all variables are initialized before use.
Don't stop at one bug.
Use debugging compilers.
Initialize constants with DATA statements or INITIAL attributes; initialize variables with executable code.
Watch out for off-by-one errors.
Take care to branch the right way on equality.
Be careful when a loop exits to the same place from side and bottom.
Make sure your code "does nothing" gracefully.
Test programs at their boundary values.
Check some answers by hand.
10.0 times 0.1 is hardly ever 1.0.
Don't compare floating point numbers solely for equality.
Make it right before you make it faster.
Make it fail-safe before you make it faster.
Make it clear before you make it faster.
Don't sacrifice clarity for small gains in "efficiency."
Let your compiler do the simple optimizations.
Don't strain to re-use code; reorganize instead.
Make sure special cases are truly special.
Keep it simple to make it faster.
Don't diddle code to make it faster—find a better algorithm.
Instrument your programs. Measure before making "efficiency" changes.
Make sure comments and code agree.
Don't just echo the code with comments—make every comment count.
Don't comment bad code—rewrite it.
Use variable names that mean something.
Use statement labels that mean something.
Format a program to help the reader understand it.
Document your data layouts.
Don't over-comment.

Appendix B
Translations of Algorithms into Pascal

This appendix demonstrates the translation of some of the algorithms and subalgorithms we have discussed into the Pascal programming language. In most of the algorithms, we did not specify what data types were to be used; we assumed the reader could determine whether a variable was numeric (integer or real), character, or boolean. The important point in algorithms is that we used each variable *consistently*—not as a character at one point and as an integer at another.

The major difference between a Pascal program and the algorithmic language we used throughout the text is that we ignored *declarations*. In some instances (like searching a list) this is beneficial: the subalgorithm we wrote is independent of the data type of the values in the array to be searched. In many of our algorithms, it doesn't matter whether the numeric variables used to store data are integer or real. They should work equally well, regardless.

Throughout this appendix, however, we are forced by the rules of Pascal to make a choice of data type for every variable. Where the choice of data type is obvious, we will use the appropriate one (e.g., counters should be integers). Where the data is numeric, but could be either real or integer, we will choose real because the program will then work for both.

Occasionally, when we declare the parameters of a procedure, we do not need to make a choice; so we will use a type name that the user of the procedure can define in a more global manner. We may need to specify certain limitations on this anonymous type. The usual limitation is that it must be a simple and not a composite type.

Each program in this appendix has been compiled and executed on the Honeywell DPS 3/E computer at the University of Kansas. Where there is a procedure without a program, it has only been compiled. Each section (below) begins by discussing points of particular interest about the choice of data type or variation from the exact specification of the algorithm. Following the discussion is a listing of the program text. Where applicable, the program is followed by a listing of a sample input list and the resulting output.

B.1. ALGORITHM 3–5 FIND THE LARGEST VALUE

This program finds the largest value in a list of values. As far as the algorithm is concerned, it does not matter whether the input list contains integer, character, or real values. For writing the algorithm in Pascal, we have chosen integers.

Another important consideration is that the read and write instructions in algorithms were defined to take/put all their parameters off of/onto one line. To obtain the same effect in Pascal, we shall use the standard procedures readln and writeln. We must also use readln because we are reading an EOF data list. The Pascal function eof can detect the end-of-file only when the end-of-line condition has been cleared. This is most easily accomplished by using readln.

Because we use readln to read in one value at a time, each value should appear on a separate line in the data file.

The output of this program consists of only one line: the line with the largest value in the input list. Stylistically, we should have written out the input list as we read it in. Because the algorithm, however, did not specify this, neither does the program.

Listing of the program

```
program FindLargest (input, output);

var
   champion,
   challenger    : integer; {this may be of any simple type}

begin

   readln (champion);          {get seeded champion value}
   while not eof do            {repeat these instructions}
      begin
         readln (challenger);
         if challenger > champion
            then
               champion := challenger {a new champion}
```

```
      end; {while}
   writeln ( 'The largest value is: ', champion:3)

end.
```

Listing of a sample data file

```
                 9
                 5
                 7
                10
                 8
                15
                19
                 3
```

Listing of the output generated

```
The largest value is: 19
```

B.2. ALGORITHM 3–6 FIND THE SMALLEST VALUE

This program, which finds the smallest value in a list of values, is nearly identical to the program in the last section. The only difference is that the check to determine whether the challenger wins over the champion is "<" rather than ">". All comments about the program to find the largest apply to this program as well.

Listing of the program

```
program FindSmallest (input, output);

var
   champion,
   challenger    : integer; {this may be of any simple type}

begin

   readln (champion);          {get seeded champion value}
   while not eof do            {repeat these instructions}
      begin
         readln (challenger);
         if challenger < champion
            then
               champion := challenger {a new champion}
```

```
      end; {while}
   writeln ( 'The smallest value is: ', champion:3)

end.
```

Listing of a sample data file

```
 9
 5
 7
10
 8
15
19
 3
```

Listing of the output generated

```
The smallest value is:  3
```

B.3. ALGORITHM 3–10 FIND THE BEST MPG

This program determines the best MPG in a list. The input list contains pairs of values on each line. The first value in the pair is the number of gallons of gas used and the second value in the pair is the number of miles traveled. Since all the data values are real values, we have declared the associated variables in the program to be real.

Listing of the program

```
program FindBestMPG (input, output);

var
   miles, gallons : real;
   BestMPG, NewMPG : real;

begin

   writeln ('The data is:');
   writeln ('Miles   Gallons   MPG');
   BestMPG := 0.0;
```

```
   while not eof do
      begin
         readln (gallons, miles);
         if (gallons > 0.0) and (miles > 0.0)
            then
               begin
                  NewMPG := miles/gallons;
                  writeln (miles:5:1, gallons:8:1, NewMPG:8:1);
                  if NewMPG > BestMPG
                     then
                        BestMPG := NewMPG
               end
            else
               begin
                  writeln (miles:5:1, gallons:8:1,
                          '---  Data must be positive.':32);
                  writeln ('These data ignored.':43)
               end
   end; {while}
if BestMPG > 0.0
   then
      writeln ('The best MPG is:', BestMPG:8:1)
   else
      writeln ('There was no data to be read.')
end.
```

Listing of a sample data file

```
   10.0   35.1
    5.1   17.0
   11.3   36.0
   -6.7   29.0
   10.5   27.2
    7.8  -48.0
    9.6   41.3
   -5.6  -16.0
    5.7   28.6
```

Listing of the output generated

```
The data is:
Miles   Gallons   MPG
 35.1    10.0     3.5
 17.0     5.1     3.3
 36.0    11.3     3.2
 29.0    -6.7     --- Data must be positive.
                      These data ignored.
 27.2    10.5     2.6
  -48     7.8     --- Data must be positive.
                      These data ignored.
 41.3     9.6     4.3
  -16    -5.6     --- Data must be positive.
                      These data ignored.
 28.6     5.7     5.0
The best MPG is:     5.0
```

B.4. ALGORITHM 3–20 PRINT THE DIGITS OF AN INTEGER

This program reads an integer value and, starting from the right-hand side, prints its digits one digit per output line. Since the algorithm uses integer arithmetic (division with truncation) to perform its task, the variables must be declared integer.

At one point in the algorithm, the remainder of dividing one integer by another is computed by the expression

number − quotient*10

in Pascal, this is more naturally expressed as

```
number mod 10
```

because the mod operator is part of the Pascal language and computes exactly the remainder we seek.

There is one subtle error in the algorithm in the text. We ignored it when we developed the algorithm because it was unimportant and would have detracted from the main discussion. If the algorithm is asked to write the digits of the integer 0—or of any negative integer—it writes out nothing. The program, because it follows the algorithm faithfully, contains the same error, as is evident by examining the output.

Listing of the program

```
program PrintDigits (input, output);

var
   number,
   quotient,
   remainder : integer;

begin

   while not eof do
      begin
         readln (number);
         writeln;
         writeln ('The number is ', number:1);
         writeln ('Its digits from right to left are:');
         while number > 0 do
            begin
               quotient := number div 10;
               remainder := number mod 10;
               writeln (remainder:3);
               number := quotient
            end {while}
      end; {while}
end.
```

Listing of a sample data file

```
        1
       27
        0
     5762
      834
     -572
   123456
```

Listing of the output generated

```
The number is 1
Its digits from right to left are:
  1
```

```
The number is 27
Its digits from right to left are:
  7
  2

The number is 0
Its digits from right to left are:

The number is 5762
Its digits from right to left are:
  2
  6
  7
  5

The number is 834
Its digits from right to left are:
  4
  3
  8

The number is -572
Its digits from right to left are:

The number is 123456
Its digits from right to left are:
  6
  5
  4
  3
  2
  1
```

B.5. ALGORITHM 4–13 COUNT A LIST OF VALUES

This program simply counts the number of values in the input list, and the two variables in this program are display and x. The first is a counter, and so should be an integer. The second could be any type that can be read using readln. Here, we chose to use real.

Listing of the program

```
program CountData (input, output);

var
   display : integer;
   x : real;      {this could be any simple type}

begin

   display := 0;
   writeln ('the data list is:');

   while not eof do
      begin
         readln (x);
         writeln (x:5:1);
         display := display + 1
      end; {while}

   writeln ('the data list contained ', display:1, ' items.')

end.
```

Listing of a sample data file

```
   10.0    27.2
    5.1     7.8
   11.3   -47.6
   -6.7     9.6
   10.5    41.3
    7.8    -5.6
    9.6   -16.3
   -5.6     5.7
    5.7    28.6
```

Listing of the output generated

```
the data list is:
 10.0
  5.1
 11.3
 -6.7
 10.5
  7.8
  9.6
 -5.6
  5.7
the data list contained 9 items.
```

B.6. ALGORITHM 4–14 SUM A LIST OF VALUES

This program sums the values in the input list. The two variables should be of the same type: the sum and the value read are both numeric and of the same data type. In this program, we chose to make both of these variables real.

Listing of the Program

```
program SumData (input, output);

var
   display,
   x           : real;          {this could be any numeric type}

begin

   display := 0.0;
   writeln ('the data list is:');

   while not eof do
      begin
         readln (x);
         writeln (x:5:1);
         display := display + x
      end; {while}

   writeln ('the data list totals ', display:6:2)

end.
```

Listing of a sample data file

```
10.0    35.1
 5.1     7.8
11.3   -47.6
-6.7     9.6
10.5    41.3
 7.8    -5.6
 7.6   -16.3
-5.6     5.7
 5.7    28.6
```

Listing of the output generated

```
the data list is:
 10.0
  5.1
 11.3
 -6.7
 10.5
  7.8
  9.6
 -5.6
  5.7
the data list totals  47.70
```

B.7. ALGORITHM 7–18 FIND THE LOWEST COST COMPUTER SYSTEM

This program finds the lowest-cost computer system, given a data list consisting of the detailed description of several computer systems. Each computer is described by a line that contains a title, followed by one line for each component that makes up the system. Each line that details a component consists of the component's name and price. The last line for each computer system follows the format of the lines that detail a component, but has 'End' for the name and 0.0 for the price.

In translating this algorithm into Pascal, we use the data type packed array of char (which is sometimes called a **string**, the usual name given this type. We use *string* (instead of *char*) so that the titles and component names may be more than one character long. This makes the program easier and more convenient (one-character names do not mean much).

We chose the real data type for the price because it's most natural. The first line of the main program initializes the variable smallest to the number 9.99e+25, which is quite a bit larger than the most expensive computer system sold (the algorithm said to be ridiculous!).

Listing of the program

```
program FindLowestCostComputer (input, output);

const
   MAXSTRING = 30;

type
   string = packed array[1 .. MAXSTRING] of char;

var
   smallest : real;
   title, SmallestTitle : string;
   totalcost : real;

procedure ProcessOneComputer (var sum:real; var title:string);

var
   component : string;
   price : real;

begin

   sum := 0.0;
   writeln;
   readln (title);
   writeln (title);
   writeln;
   readln (component, price);
   while component <> 'End' do
      begin
         writeln ('    ', component, price:8:2);
         sum := sum + price;
         readln (component, price)
      end; {while}
   writeln;
   writeln ('   Total system cost is: ', sum:8:2);

end; {ProcessOneComputer}

begin {Program FindLowestCostComputer}
```

```
   smallest := 9.99e+25;  {no system could cost this much!}
   SmallestTitle := 'There was no data to be read.';

   while not eof do
      begin
         ProcessOneComputer (totalcost, title);
         if totalcost < smallest
            then
               begin
                  smallest := totalcost;
                  SmallestTitle := title
               end
      end; {while}

   writeln ('The smallest total system cost is: ',
                    smallest:8:2);
   writeln ('The system with the smallest total cost is: ',
                         SmallestTitle)

end.
```

Listing of a sample data file

```
Computer Shack Ziggy-V
   Basic-Unit                  670.0
   CRT                          80.0
   16K-memory                  125.0
   OMNICALC                    175.0
   Printer                     215.0
   End                           0.0

Itty Bitty Machines Pico-I
   Basic-Unit                  500.0
   CRT                          95.0
   64K-memory                  350.0
   QUASICALC                    25.0
   Printer                     189.95
   End                            .0
```

```
Sushiyama All-In-One
   Basic-Unit                     995.0
   CRT-included                     0.0
   Printer-included                 0.0
   SUSHICALC                        0.0
   End                              0.0
```

Listing of the output generated

```
Computer Shack Ziggy-V
   Basic-Unit                         670.00
   CRT                                 80.00
   16K-memory                         125.00
   OMNICALC                           175.00
   Printer                            215.00

   Total system cost is:  1265.00

Itty Bitty Machines Pico-I
   Basic-Unit                         500.00
   CRT                                 95.00
   64K-memory                         350.00
   QUASICALC                           25.00
   Printer                            189.95

   Total system cost is:  1159.95

Sushiyama All-In-One
   Basic-Unit                         995.00
   CRT-included                         0.
   Printer-included                     0.
   SUSHICALC                            0.

   Total system cost is:   995.00
The smallest total system cost is:   995.00
The system with the smallest total cost is: Sushiyama All-In-Or
```

B.8. ALGORITHM 8–10 REVERSE A LIST

In this program, which writes out the input list in reverse, several data types are declared. ElementType is the type of each element of the array. For this program, we have set ElementType to be the same as the Pascal char data type; so the input list this program expects is a number of lines consisting of a single character each. (Also, we could have set ElementType to integer or real to reverse lists of those types of values.)

The data type index is used to declare variables (like the placekeeper) that hold index values in the array. It is declared to be the subrange of integers, starting at zero and ending at MAXINDEX. MAXINDEX was declared to be a constant identifier, equal to 100 in this program. The type index begins at zero because variables of this type may have to be set to zero to indicate that the array is empty. ListType is the name of the type of the array we will use to store the individual values from the input list. The index type of the array declared in ListType is a subrange of integers, beginning at one and ending at MAXINDEX to coincide with everything but zero in the subrange declared in index. The index type of the array does not include zero as an index, however, because we reserve that to mean an empty array.

This program does not check for array overflow (i.e., the number of data lines does not exceed the number of cells in the array). If the input list happened to have more than MAXINDEX lines (equal to 100 here), the program would cause an error. Also, the algorithm did not specify that a heading be written out before the reversed input list. This was added in the program.

Listing of the program

```
program ReverseList (input, output);

const
   MAXINDEX = 100;

type
   index = 0 .. MAXINDEX;
   ElementType = char;
   ListType = array [1 .. MAXINDEX] of ElementType;

var
   count,
   i : index;
   X : ListType;

begin

   count := 0;
   while not eof do   {read the list into the array}
      begin
         count := count + 1;
         i := count;
```

```
        readln (X[i])
     end; {while}

  writeln ('The reverse of the input list is:');
  writeln;
  for i := count downto 1 do    {write out the list backwards}
     writeln (X[i]:5);

  if count = 0
     then
        writeln ('The data list is empty.')
end.
```

Listing of a sample data file

```
H
e
l
l
o
!
```

Listing of the output generated

```
The reverse of the input list is:
    !
    o
    l
    l
    e
    H
```

B.9. ALGORITHM 8–14 READ AN EOF LIST INTO AN ARRAY

This procedure reads the input list to its end-of-file and stores each data item into consecutive elements of an array, beginning at element index 1. To demonstrate this procedure, we chose the integer type. The input list is assumed to be a list of integers, arranged one to a line. To make the procedure work for any *readable* type, change the type ElementType to that type. (Remember: the readable types are integer, real, and char.) As in the last program, the types index and ListType are declared for the placekeeper and the array respectively.

Listing of the procedure

```
{For purposes of this example, we will assume the types
 'ListType' and 'index' are defined as below. In a real
 application, you should change them to what is appropriate.
}
const
   MAXINDEX = 100;

type
   index = 0 .. MAXINDEX;
   ElementType = integer;
   ListType = array [1 .. MAXINDEX] of ElementType;

procedure ReadEOFList (var X : ListType; var COUNT : index);

var
   value : ElementType;

begin

   COUNT := 0;
   while (COUNT < MAXINDEX) and (not eof) do
      begin
         readln (value);
         COUNT := COUNT + 1;
         X[COUNT] := value
      end; {while}

   if not eof
      then
         begin
            writeln ('The data list is too long for this',
                              ' program.');
            writeln ('Found by: read an eof list.')
         end

end; {procedure ReadEOFList}
```

B.10. ALGORITHM 8–15 READ A SENTINEL LIST INTO AN ARRAY

This procedure reads the input list until it encounters the sentinel value, and stores each data item into consecutive elements of an array, starting at element index 1. We assume that there is a unique sentinel value for translating this subalgorithm. In some cases, if there is a range of sentinel values (e.g., any negative number), it is necessary to modify the procedure to stop reading when it encounters such a sentinel value. Except that the procedure reads in a sentinel list, it is the same as the procedure in the last section, which reads in an end-of-file list.

Listing of the procedure

```
{For purposes of this example, we will assume the types
 'ListType' and 'index' are defined as below. In a real
 application, you should change them to what is appropriate.
 The constant SENTINEL should also be changed as appropriate.
}

const
   MAXINDEX = 100;
   SENTINEL = -1;

type
   index = 0 .. MAXINDEX;
   ElementType = integer;
   ListType = array [1 .. MAXINDEX] of ElementType;

procedure ReadSentinelList (var X:ListType; var COUNT:index);

var
   value : ElementType;

begin

   COUNT := 0;
   readln (value);
   while (COUNT < MAXINDEX) and (value <> SENTINEL) do
      begin
         COUNT := COUNT + 1;
         X[COUNT] := value;
         readln (value)
      end; {while}

   if value <> SENTINEL
      then
         begin
            writeln ('The data list is too long for this
                              ' program.');
```

```
            writeln ('Found by: read a sentinel list.')
         end

end; {procedure ReadSentinelList}
```

B.11. ALGORITHM 8–16 READ A HEADER LIST INTO AN ARRAY

This procedure reads a header type input list into an array. As in the last two sections, the three types ElementType, index, and ListType are declared. The variable named header is declared to be an integer because it is a counting variable. It was *not* declared to be type index because we could not perform the error checks ourselves. That is, if a bad value were read in, the Pascal support system might intervene and cause an error before we could check for ourselves.

Listing of the procedure

```
{For purposes of this example, we will assume the types 'list'
 and 'index' are defined as below. In a real application, you
 should change them to what is appropriate.
}
const
   MAXINDEX = 100;

type
   index = 0 .. MAXINDEX;
   ElementType = integer;
   list = array [1 .. MAXINDEX] of ElementType;

procedure ReadHeaderList (var X : list; var COUNT : index);

var
   value : ElementType;
   header : integer;

begin

   readln (header);
   if header > MAXINDEX
      then
         begin
```

```
            writeln ('The data list is too long for this');
            writeln ('program. Only the first ', MAXINDEX:1,
                     ' values will be read.');
            writeln (' Found by: read a header list.');
            header := MAXINDEX
         end;

   for COUNT := 1 to header do
      begin
         readln (value);
         X[COUNT]:= value
      end; {for}

   COUNT := header

end; {procedure ReadHeaderList}
```

B.12. ALGORITHM 8–20 SUM THE VALUES IN AN ARRAY

This procedure sums the values in the array it is given as a precondition. Again, the types ElementType, index, and ListType are declared to allow us to declare variables and parameters. In this procedure, ElementType could be either integer or real, as needed. We have chosen to use integer. The sum is declared to be ElementType because the sum of integers is an integer, and likewise for reals. Thus the type of the sum should be the same as the type of the elements of the array.

Listing of the procedure

```
{For purposes of this example, we will assume the types
 'ListType' and 'index' are defined as below. In a real
 application, you should change them to what is appropriate.
}

const
   MAXINDEX = 100;

type
   index = 0 .. MAXINDEX;
   ElementType = integer; {This could be either  integer or rea
   ListType = array [1 .. MAXINDEX] of ElementType;
```

```
procedure SumArray (X:ListType; N:index; var sum:ElementType);

var
   i : index;

begin

   sum := 0;
   for i := 1 to N do
      sum := sum +X[i]

end; {procedure SumArray}
```

B.13. ALGORITHM 8–21 PROGRAM TO AVERAGE A LIST OF VALUES

This program averages the values in a sentinel input list. The algorithm referred to in the section heading is only a subalgorithm to average an array, but this provides a good opportunity to show a program with multiple procedures. We use the procedures presented in previous sections to read a sentinel input list into an array and to sum the values in an array. The subalgorithm to average an array is given here as a procedure as well. The main program calls upon the needed procedures and does some error checking, as well as writing out the input list and the result.

Note that our consistent use of the type names ElementType, index, and ListType, together with the constant name MAXINDEX has made it easy to collect these previously written procedures. It was not necessary to change any of the procedure because all of them used common type names.

Listing of the program

```
program AverageValues (input, output);

const
   MAXINDEX = 100;
   SENTINEL = -1;

type
   index = 0 .. MAXINDEX;
   ElementType = real;    {this may be either real or integer}
   ListType = array [1 .. MAXINDEX] of ElementType;
```

```
var
   i : index;
   Length : index;
   ValueList : ListType;
   Avg : real;

procedure SumArray (X:ListType; N:index; var sum:ElementType);

var
   i : index;

begin

   sum := 0;
   for i := 1 to N do
      sum := sum +X[i]

end; {procedure SumArray}

procedure AverageArray (LIST : ListType; N : index;
                        var AVERAGE : real);

var
   TOTAL : ElementType;

begin

   if N > 0
      then
         begin
            SumArray (LIST, N, TOTAL);
            AVERAGE := TOTAL / N
         end
      else
         begin
            AVERAGE := 0.0;
            writeln ('An attempt to average an empty list);
            writeln ('was detected in AverageArray.')
         end

end; {AverageArray}

procedure ReadSentinelList (var X:ListType; var COUNT:index);
```

```
var
   value : ElementType;

begin

   COUNT := 0;
   readln (value);
   while (COUNT < MAXINDEX) and (value <> SENTINEL) do
      begin
         COUNT := COUNT +1;
         X[COUNT]:= value;
         readln (value)
      end; {while}

   if value <> SENTINEL
      then
         begin
            writeln ('The data list is too long for this',
                        'program.');
            writeln ('Found by: read a sentinel list.')
         end

end; {procedure ReadSentinelList}

begin {Program AverageList}

   ReadSentinelList (ValueList, Length);
   if Length <= 0
      then
         writeln ('The data list is empty.')
      else
         begin
            writeln ('The input list is:');
            writeln;
            for i := 1 to Length do
               writeln (ValueList[i]:10:2);
            writeln;

            AverageArray (ValueList, Length, Avg);
            writeln ('The average of these values is: ', Avg:8:2)
         end
end.
```

Listing of a sample data file

```
 76
 24
-15
 27
  5
  0
 16
 21
  8
 -1
```

Listing of the output generated

```
The input list is:
      76.00
      24.00
     -15.00
      27.00
       5.00
       0.
      16.00
      21.00
       8.00
The average of these values is:     18.00
```

B.14. ALGORITHM 9–5 SEARCH A LIST FOR A PARTICULAR VALUE

This procedure implements the search algorithm. Again, we use the common type and constant names we have used for array and placekeeper declarations. ElementType could be any type whose values can be compared for equality by using the " = " relational operator. Y is declared to be of ElementType because it and all the values in the array must be of the same type. Y, after all, is to be compared with elements in X.

Listing of the Procedure

```
{For purposes of this example, we will assume the types
 'ListType' and 'index' are defined as below. In a real
 application, you should change them to what is appropriate.
}
```

```
const
   MAXINDEX = 100;

type
   index = 0 .. MAXINDEX;
   ElementType = integer;
   ListType = array [1 .. MAXINDEX] of ElementType;

procedure Search (X : ListType; PlaceKeeper : index;
                  Y : ElementType; var found : boolean;
                  var position : index);

var
   scan : index;

begin

   found := false;
   scan := 0;
   while (scan < PlaceKeeper) and (not found) do
      begin
         scan := scan + 1;
         if X[scan] = Y
            then
               found := true
      end; {while}
   position := scan

end; {Search}
```

B.15. ALGORITHM 9–9 USING PARALLEL ARRAYS, PRINT FREQUENCIES

This program constructs a list of values from the input list in such a way that only one instance of an input value occurs in an array. It also counts the number of times each value occurs. (This was described in discussion of the algorithm as building a list of unique values and frequencies.) This implementation uses parallel arrays. In the next section, we will translate the algorithm by using an array of records.

Rather than simply present a procedure (the natural translation of the subalgorithm referred to in the section heading), we present a program that uses such a procedure. We also use the search procedure translated in the last section.

In this problem, we need two distinct arrays: one to hold the individual values and another to hold the count of the occurrences of the corresponding values. In this case, ElementType applies only to the array that holds the values read from the input list. The variables that are declared of type ListFreq have an element type of integer (each element is a counter). The variables that are declared of type ListValues hold the values read in. ListValues could be any type whose values can be compared by using the relational operator " = ".

Procedure Search must be given before procedure ConstructUnique, because Search is called by ConstructUnique.

Listing of the program

```
program PrintFrequencies (input, output);

const
   MAXINDEX = 250;

type
   index = 0 .. MAXINDEX;
   ElementType = integer;
   ListValues = array [1 .. MAXINDEX] of ElementType;
   ListFreq = array [1 .. MAXINDEX]of integer;

var
   Length : index;
   UniqueValues : ListValues;
   Frequencies : ListFreq;
   i : index;

procedure Search (X : ListValues; PlaceKeeper : index;
                  Y : ElementType; var found : boolean;
                  var position : index);

var
   scan : index;

begin

   found := false;
   scan := 0;
   while (scan < PlaceKeeper) and (not found) do
      begin
         scan := scan + 1;
         if X[scan] = Y
```

```
          then
             found := true
      end; {while}
   position := scan

end; {Search}

procedure ConstructUnique (var unique : ListValues;
                           var frequency : ListFreq;
                           var size : index);

var
   position : index;
   value : ElementType;
   found : boolean;

begin

   size := 0;
   while (size < MAXINDEX) and (not eof) do
      begin
         readln (value);
         Search (unique, size, value, found, position);
         if not found
            then
               begin
                  size := size + 1;
                  unique[size] := value;
                  frequency[size] := 1
               end
            else
               frequency[position] := frequency[position] +1
      end; {while}

   if not eof
      then
         begin
            writeln ('The data list is too long for this',
                          'program.');
            writeln ('Found by: ConstructUnique.')
         end;

end; {ConstructUnique}
```

```
begin {program PrintFrequencies}

   ConstructUnique (UniqueValues, Frequencies, Length);

   writeln ('The unique values in the data and their frequenc
   writeln;
   writeln ('       value       frequency')
   writeln;

   for i := 1 to Length do
      writeln (UniqueValues[i]:11, Frequencies[i]:11);

end. {program PrintFrequencies}
```

Listing of a sample data file

```
 21
 17
 21
 16
 17
 21
 10
 16
 17
 10
  5
  5
 10
  5
 17
 17
  8
 16
100
  5
  8
```

Listing of the output generated

```
The unique values in the data and their frequency

      value       frequency
        21            3
        17            5
        16            3
        10            3
         5            4
         8            2
       100            1
```

B.16. ALGORITHM 9–10 USING RECORDS, PRINT FREQUENCIES

This program implements the same algorithm as the program in the previous section, except that this algorithm uses a single array whose element type is a record with two fields (rather than two parallel arrays).

ElementType in this program is a record consisting of a field named data to hold the data value, and frequency to hold the count of occurrences of that value. The new type name, DataType, has been introduced to allow us to refer uniformly to the type of the values on the input list and, to the type that should be used to declare the data field within the record.

Listing of the program

```
program PrintFrequencies (input, output);

const
   MAXINDEX = 250;

type
   index = 0 .. MAXINDEX;
   DataType = integer;
   ElementType = record
                    data : DataType;
                    frequency : integer
                 end;
   ListType = array [1 .. MAXINDEX] of ElementType;
```

```
var
   Length : index;
   UniqueValues : ListType;
   i : index;

procedure Search (X:ListType; PlaceKeeper:index; Y:DataType;
                  var found : boolean; var position : index);

var
   scan : index;

begin

   found := false;
   scan := 0;
   while (scan < PlaceKeeper) and (not found) do
      begin
         scan := scan + 1;
         if X[scan].data = Y
            then
               found := true
      end; {while}
   position := scan

end; {Search}

procedure ConstructUnique (var list:ListType; var size:index);

var
   position : index;
   value : DataType;
   found : boolean;

begin

   size := 0;
   while (size < MAXINDEX) and (not eof) do
      begin
         readln (value);
         Search (list, size, value, found, position);
         if not found
            then
               begin
                  size := size + 1;
                  list[size].data := value;
```

```
                list[size].frequency := 1
             end
          else
             list[position].frequency :=
                      list[position].frequency +1
      end; {while}

   if not eof
      then
         begin
            writeln ('The data list is too long for this',
                        'program.');
            writeln ('Found by: ConstructUnique')
         end;

end; {ConstructUnique}

begin   {program PrintFrequencies}

   ConstructUnique (UniqueValues, Length);
   writeln ('The unique values and their frequency');
   writeln;
   writeln ('      value     frequency');
   writeln;

   for i := 1 to Length do
      writeln (UniqueValues[i].data:11,
               UniqueValues[i].frequency:11);

end. {program PrintFrequencies}
```

Listing of a sample data file

```
        21
        17
        21
        16
        17
        21
        10
        16
        17
        10
         5
         5
        10
         5
```

```
17
17
 8
16
100
 5
 8
```

Listing of the output generated

```
The unique values in the data and their frequency

      value    frequency
        21         3
        17         5
        16         3
        10         3
         5         4
         8         2
       100         1
```

B.17. ALGORITHM 9–13 PROGRAM TO SORT AN ARRAY

This program reads in an end-of-file input list and sorts it into ascending sequence. The subalgorithm referred to in the section heading is the insertion sort subalgorithm. We have translated that subalgorithm into the procedure Sort and included the required procedures to FindSmallest and switch, along with ReadEOFList. We wrote a main program to call the reading and sorting procedures and to write out the resulting sorted array.

Listing of the program

```
program SortDataList (input, output);

const
    MAXINDEX = 100;
type
   index = 0 .. MAXINDEX;
   ElementType = integer;  {may be any simple type}
   ListType = array [1 .. MAXINDEX] of ElementType;
```

```
var
   DataList : ListType;
   Length : index;
   i : index;

procedure switch (var X, Y : ElementType);

var
   temp : ElementType;

begin

   temp := X;
   X := Y;
   Y := temp

end; {procedure switch}

procedure FindSmallest (X:ListType; N, K:index; var p:index);

var
   i : index;

begin

   p := K;
   for i := K to N do
      if X[i] < X[p]
         then
            p := i

end; {procedure FindSmallest}

procedure Sort (var X : ListType; size : index);

var
   j, p : index;

begin

   for j := 1 to size-1 do
      begin
         FindSmallest (X, size, j, p);
         switch (X[j], X[p]);
      end {for}
```

```
end; {procedure sort}

procedure ReadEOFList (var X : ListType; var COUNT : index);

var
   value : ElementType;

begin

   COUNT := 0;
   while (COUNT < MAXINDEX) and (not eof) do
      begin
         readln (value);
         COUNT := COUNT + 1;
         X[COUNT] := value
      end; {while}

   if not eof
      then
         begin
            writeln ('The data list is too long for this',
                               ' program.');
            writeln ('Found by: read an eof list.')
         end

end; {procedure ReadEOFList}

begin {SortDataList}

   ReadEOFList (DataList, Length);
   Sort (DataList, Length);
   writeln ('The sorted data list is:');
   writeln;
   for i := 1 to Length do
      writeln (Datalist[i]:5)

end.
```

Listing of a sample data file

```
122
 76
 50
 81
 10
 36
  5
  0
110
 75
 81
  6
 43
```

Listing of the output generated

```
The sorted data list is:
     0
     5
     6
    10
    36
    43
    50
    75
    76
    81
    81
   110
   122
```

Appendix C Translations of algorithms into FORTRAN 77

In this appendix we demonstrate the translation of some of the algorithms and subalgorithms we have discussed into the FORTRAN 77 programming language. In most of the algorithms, we did not specify what data types were to be used, but assumed the reader could determine whether a variable is numeric (integer or real), character, or logical. The important point to be aware of in algorithms is that we use each variable *consistently*, not as a character at one point and as an integer at another.

The major difference between a FORTRAN 77 program and the algorithmic language the text is that we ignored declarations. In some instances (like searching a list) this is beneficial: the subalgorithm we wrote is independent of the particular data type of the values in the array to be searched. In many of our algorithms it doesn't matter whether the numeric variables used to store data are integer or real. They should work equally well regardless.

Throughout this appendix, however, we are forced by the rules of FORTRAN 77 to make a choice of data type for every variable. Where the choice of data type is obvious, we will use the appropriate one (counters should be integers, for example). Where the data is numeric, but could be either real or integer, we will choose real because the program will work for both.

Each program in this appendix has been compiled and executed on the Honeywell DPS 3/E computer at the University of Kansas. Where there is only a subroutine, it has only been compiled. Each of the sections below begins by discussing points of particular interest about the choice of data type or variation from the specification of the algorithm. Following the discussion is a listing of the program text. Where applicable, the program is followed by a listing of a sample input list and the resulting output.

Whereas we used names of whatever length we liked in algorithms, most FORTRAN 77 compilers limit variable names to a maximum of six characters. In the translation, we have abbreviated long names so that they might still have some meaning to the reader.

Since there is no while loop in F77, we will translate:

```
while condition do        100 if (condition) then
   statements         to          statements
end while                         goto 100
                              end if
```

ALGORITHM EXAMPLE C–1

This translation works fine, except when we translate end-of-file read loops. We cannot follow the algorithm exactly because there is no way to detect the end of a file in f77 in a "condition." We can only check for the end of a file by trying to read from that file and using the "END = " option of the read statement. In these cases, we translate the while loop as

```
while there is data do     100 continue
   read, value                    read (*, *, end=110) value
   statements         to          statements
end while                         goto 100
                           110 continue
```

ALGORITHM EXAMPLE C–2

This is not very pretty, but it is the most straightforward way to translate the loop.

C.1. ALGORITHM 3–5 FIND THE LARGEST VALUE

This program finds the largest value in a list of values. As far as the algorithm is concerned, it does not matter whether the input list contains integer, character, or real values. For purposes of writing the algorithm in FORTRAN 77, we have chosen to use integers.

Because we use read to read in one value at a time, each value should appear on a separate line in the data file.

The output of this program consists only of one line: the line giving the largest value in the input list. Stylistically, we should have written out the input list as we read it in. The algorithm, however, did not specify this, so neither does the program.

Listing of the program

```
      program FindLg
c
      integer champ, chall
c
      read *, champ
c
  100 continue
         read (*, *, end=110) chall
         if (chall .gt. champ) then
            champ = chall
         end if
      goto 100
  110 continue
c
      print *, 'The largest value is: ', champ
c
      end
```

Listing of a sample data file

```
         9
         5
         7
        10
         8
        15
        19
         3
```

Listing of the output generated

```
The largest value is:              19
```

C.2. ALGORITHM 3–6 FIND THE SMALLEST VALUE

This program finds the smallest value in a list of values. It is nearly identical to the program in the last section. The only difference is that the check to determine whether the challenger wins over the champion is ".lt." rather than ".gt.". All of the comments about the program to find the largest apply to this program as well.

Listing of the program

```
      program FindSm
c
      integer champ, chall
c
      read *, champ
c
  100 continue
         read (*, *, end=110) chall
         if (chall .lt. champ) then
            champ = chall
         end if
      goto 100
  110 continue
c
      print *, 'The smallest value is: ', champ
c
      end
```

Listing of a sample data file

```
 9
 5
 7
10
 8
15
19
 3
```

Listing of the output generated

```
The smallest value is:                3
```

C.3. ALGORITHM 3–10 FIND THE BEST MPG

This program determines the best MPG in a list. The input list contains pairs of values on each line. The first value in the pair is the number of gallons of gas used and the second value in the pair is the number of miles traveled. Since all of the data values are real values, we have declared the associated variables in the program to be real.

Listing of the program

```
      program FndMPG
c
      real miles, gals
      real BstMPG, NewMPG
c
      print *, 'The data is:'
      print *, 'Miles   Gallons   MPG'
      BstMPG = 0.0
c
  100 continue
         read (*, *, end=110) gals, miles
         if ( (gals .gt. 0.0) .and. (miles .gt. 0.0) ) then
            NewMPG = miles/gals
            print *, miles, gals, NewMPG
            if (NewMPG .gt. BstMPG) then
               BstMPG = NewMPG
            end if
         else
            print *, miles, gals, '        --- Data must be positive.'
            print *, '                         These data ignored.'
         end if
      goto 100
  110 continue
c
      if (BstMPG .gt. 0.0) then
         print *, 'The best MPG is: ', BstMPG
      else
         print *, 'There was no data to be read.'
      end if
c
      end
```

Listing of a sample data file

```
10.0   35.1
 5.1   17.0
11.3   36.0
-6.7   29.0
10.5   27.2
 7.8  -47.6
 9.6   41.3
-5.6  -16.3
 5.7   28.6
```

Listing of the output generated

```
The data is:

          Miles          Gallons        MPG
      35.0999999     10.0000000    3.5100000
      17.0000000      5.1000000    3.3333333
      36.0000000     11.3000000    3.1858407
      29.0000000     -6.7000000    --- Data must be positive.
                       These data ignored.
      27.2000000     10.5000000    2.5904762
     -47.5999999      7.8000000    --- Data must be positive.
                       These data ignored.
      41.3000002      9.6000000    4.3020833
     -16.3000000     -5.6000000    --- Data must be positive.
                       These data ignored.
      28.5999999      5.7000000    5.0175439
The best MPG is:      5.0175439
```

C.4. ALGORITHM 3–20 PRINT THE DIGITS OF AN INTEGER

This program reads an integer value and, starting from the right hand side, prints its digits one digit per output line. Since the algorithm uses integer arithmetic (division with truncation) to perform its task, the variables must necessarily be declared integer.

There is one subtle error in the algorithm given in the text. We ignored it at the time we developed the algorithm because it was unimportant and would have distracted from the main discussion. If the algorithm is asked to write the digits of the integer 0, or of any negative integer, it writes out nothing instead. The program, because it follows the algorithm faithfully, also contains the same error as is evident by examining the output produced.

Listing of the program

```
      program Digits
c
      integer number, quotnt, remain
c
  100 continue
         read (*, *, end=120) number
         print *,
         print *, 'The number is ', number
         print *, 'Its digits from right to left are:'
```

```
c
  110     if (number .gt. 0) then
             quotnt = number / 10
             remain = number - quotnt*10
             print *, remain
             number = quotnt
             goto 110
          end if
c
      goto 100
  120 continue
c
      end
```

Listing of a sample data file

```
            1
           27
            0
         5762
          834
         -572
       123456
```

Listing of the output generated

```
The number is                 1
Its digits from right to left are:
               1
The number is                27
Its digits from right to left are:
               7
               2

The number is                 0
Its digits from right to left are:

The number is              5762
Its digits from right to left are:
               2
               6
               7
               5
```

```
The number is                 834
Its digits from right to left are:
              4
              3
              8

The number is                -572
Its digits from right to left are:

The number is              123456
Its digits from right to left are:
              6
              5
              4
              3
              2
              1
```

C.5. ALGORITHM 4–13 COUNT A LIST OF VALUES

This program simply counts the number of values in the input list. The two variables in this program are displa and x. The first is a counter, and so should be an integer. The second could be any type. Here, we chose to use real.

Listing of the program

```
      program Count
c
      integer displa
      real x
c
      displa = 0
      print *, 'the data list is:'
c
  100 continue
         read (*, *, end=110) x
         print *, x
         displa = displa + 1
      goto 100
  110 continue
c
      print *, 'the data list contained ', displa, ' items.'
      end
```

Listing of a sample data file

```
    10.0    35.1
     5.1    17.0
    11.3    36.0
    -6.7    29.0
    10.5    27.2
     7.8   -47.6
     9.6    41.3
    -5.6   -16.3
     5.7    28.6
```

Listing of the output generated

```
the data list is:
      10.0000000
       5.1000000
      11.3000000
      -6.7000000
      10.5000000
       7.8000000
       9.6000000
      -5.6000000
       5.7000000
the data list contained            9 items.
```

C.6. ALGORITHM 4–14 SUM A LIST OF VALUES

This program sums the values in the input list. The two variables should be of the same type: the sum and the value read are both numeric and of the same data type. In this program we chose to make both of these variables real.

Listing of the program

```
      program SumDat
c
      real displa, x
c
      displa = 0.0
```

```
      print *, 'the data list is:'
c
  100 continue
         read (*, *, end=110) x
         print *, x
         displa = displa + x
      goto 100
  110 continue
c
      print *, 'the data list totals ', displa
      end
```

Listing of a sample data file

```
     10.0      35.1
      5.1      17.0
     11.3      36.0
     -6.7      29.0
     10.5      27.2
      7.8     -47.6
      9.6       1.3
     -5.6     -16.3
      5.7      28.6
```

Listing of the output generated

```
the data list is:
      10.0000000
       5.1000000
      11.3000000
      -6.7000000
      10.5000000
       7.8000000
       9.6000000
      -5.6000000
       5.7000000
the data list totals            47.6999998
```

C.7. ALGORITHM 7–18 FIND THE LOWEST COST COMPUTER SYSTEM

This program finds the lowest-cost computer system, given a data list consisting of the detailed description of several systems. Each computer is described by a line containing a title, followed by one line for each component making up the system. Each line detailing a component consists of the component's name and price. The last line for each computer system follows the format of the lines detailing a component, but has 'End' for the name and 0.0 for the price.

The titles and component names in the input list could be quite long. We have declared these variables to be 30 characters each. We chose the real data type for the price because that is the most natural. The first line of the main program initializes the variable smallest to the number 9.99e+25, which is quite a bit larger than the most expensive computer system sold (the algorithm said to be ridiculous!)

We use formatted reads and prints in this program as an example because the free format output looked particularly bad. Normally, we use free format input and output so we need not be distracted by the tedium of formatting.

Listing of the program

```
      program FndLow
c
      real small, total
      character*30 title, Stitle
      logical eofile
      external eofile
c
      small = 9.99e+25
      Stitle = 'There was no data to be read.'
c
  100 if (.not. eofile (0)) then
         call Procl (total, title)
         if (total .lt. small) then
            small = total
            Stitle = title
         end if
      goto 100
      end if
c
      print 900, small
      print 910, Stitle
c
  900 format(' The smallest total system cost is: ', f8.2)
  910 format(' The system with the smallest total cost is: ',a30)
      end
c
c
c
      subroutine Procl (sum, title)
c
```

```
      real sum, price
      character*30 title, compon
c
      sum = 0.0
      print *,
      read 900, title
      print 910, title
      print *,
      read 930, compon, price
c
  100 if (compon .ne. 'End') then
         print 920, compon, price
         sum = sum + price
         read 930, compon, price
         goto 100
      end if
c
      print *,
      print 940, sum
c
  900 format (a30)
  910 format (1x, a30)
  920 format (4x, a30, f8.2)
  930 format (a30, f10.0)
  940 format ('    Total system cost is: ', f8.2)
        end
c
c
c
c   This function is required to detect the end of file in the
c   main program FNDLOW because all of the reading is done
c   in the subroutine PROC1. Thus to opportunity to detect
c   the end of file via the END= option is unavailable in
c   the main program.
c
c
      logical function eofile (unit)
c
      integer unit, intern
c
      if (unit .eq. 0) then
c
c             set intern to standard read unit number

         intern = 41
      else if (unit .gt. 0) then
         intern = unit
```

```
      else
         print *, 'Illegal unit number ', unit, ' in EOFILE.'
         return
      end if
      read (intern, *, end=99)
c
      eofile = .false.
      backspace intern
      return
c
   99 continue
      eofile = .true.
      backspace intern
c
      end
```

Listing of a sample data file

```
Computer Shack Ziggy-V
   Basic-Unit                    670.0
   CRT                            80.0
   16K-memory                    125.0
   OMNICALC                      175.0
   Printer                       215.0
   End                             0.0
Itty Bitty Machines Pico-I
   Basic-Unit                    500.0
   CRT                            95.0
   64K-memory                    350.0
   QUASICALC                      25.0
   Printer                      189.95
   End                             0.0
Sushiyama All-In-One
   Basic-Unit                    995.0
   CRT-included                    0.0
   Printer-included                0.0
   SUSHICALC                       0.0
   End                             0.0
```

Listing of the output generated

```
Computer Shack Ziggy-V

   Basic-Unit                         670.00
   CRT                                 80.00
   16K-memory                         125.00
   OMNICALC                           175.00
   Printer                            215.00

   Total system cost is:  1265.00

Itty Bitty Machines Pico-I

   Basic-Unit                         500.00
   CRT                                 95.00
   64K-memory                         350.00
   QUASICALC                           25.00
   Printer                            189.90

   Total system cost is:  1159.90

Sushiyama All-In-One

   Basic-Unit                         995.00
   CRT-included                         0.00
   Printer-included                     0.00
   SUSHICALC                            0.00

   Total system cost is:   995.00
The smallest total system cost is:    995.00
The system with the smallest total cost is: Sushiyama All-In-C
```

C.8. ALGORITHM 8–10 REVERSE A LIST

In this program (which writes out the input list in reverse) we have chosen to have the array X contain one character per element, so the input list this program expects is a number of lines consisting of a single character each. Also, we could have set the array element type to integer or real to reverse lists of those types of values. Whichever type we choose, both the array X and the variable value should have the same type. MAXINDEX was declared to be a symbolic constant, equal to 100 in this program.

This program does not check for array overflow (the number of data lines does not exceed the number of cells in the array). If the input list had more than MAXINDEX lines (equal to 100 here), the program would cause an error. Also, the algorithm did not specify that a heading be written out before the reversed input list. This was added in the program.

Listing of the program

```
      program Revers
c
      parameter (MAXIDX = 100)
      integer count, i
      character*1 X(MAXIDX), value
c
      count = 0
c
  100 continue
         read (*, *, end=110) value
         count = count +1
         i = count
         X(i) = value
         goto 100
  110 continue
c
      print *, 'The reverse of the input list is:'
      print *,
c
      do 120, i = count, 1, -1
         print *, X(i)
  120 continue
c
      if (count .eq. 0) then
         print *, 'The data list is empty.'
      end if
c
      end
```

Listing of a sample data file

```
           H
           e
           l
           l
           o
           !
```

Listing of the output generated

```
 The reverse of the input list is:
 !
 o
 l
 l
 e
 H
```

C.9. ALGORITHM 8–14 READ AN EOF LIST INTO AN ARRAY

This subroutine reads the input list to its end-of-file and stores each data item into consecutive elements of an array, beginning at element index 1. To demonstrate this subroutine, we chose the integer type. The input list is assumed to be a list of integers, arranged one to a line. To make the subroutine work for any type, change the declarations of both X and value to that type.

Listing of the subroutine

```
      subroutine RdEOF (X, COUNT)
c
      parameter (MAXIDX = 100)
      integer COUNT
      integer value
      integer X(*)
c
      COUNT = 0
c
  100 if (COUNT .lt. MAXIDX) then
         read (*, *, end=110) value
         COUNT = COUNT + 1
         X(COUNT) = value
         goto 100
      end if
c
      print *, 'The data list is too long for this program.'
      print *, 'This error was found by: read an eof list.'
c
  110 continue
c
      end
```

C.10. ALGORITHM 8–15 READ A SENTINEL LIST INTO AN ARRAY

This subroutine reads the input list until it encounters the sentinel value, then stores each data item into consecutive elements of an array, starting at element index 1. We assume there is a unique sentinel value for purposes of translating this subalgorithm. If there is a range of sentinel values (e.g., any negative number), it is necessary to modify the subroutine to stop reading when it encounters such a value. Except that the subroutine reads in a sentinel list, it is the same as the subroutine (in the last section) that reads in an end-of-file list.

Listing of the subroutine

```
      subroutine RdSent (X, COUNT)
c
      parameter (MAXIDX = 100)
      parameter (SENTNL = -1)
      integer COUNT
      integer value
      integer X(*)
c
      COUNT = 0
      read *, value
c
  100 if ((COUNT .lt. MAXIDX) .and. (value .ne. SENTNL)) then
         COUNT = COUNT +1
         X(COUNT) = value
         read *, value
         goto 100
      end if
c
      if (value .ne. SENTNL) then
         print *, 'The data list is too long for this program.'
         print *, 'Found by: read a sentinel list.'
      end if
c
      end
```

C.11. ALGORITHM 8–16 READ A HEADER LIST INTO AN ARRAY

In this subroutine, which reads a header-type input list into an array, the variable named header is declared to be an integer because it is a counting variable.

Listing of the subroutine

```
      subroutine RdHead (X, COUNT)
c
      parameter (MAXIDX = 100)
      integer COUNT, header
      integer value
      integer X(*)
c
      read *, header
      if (header .gt. MAXIDX) then
         print *, 'The data list is too long for this program.
         print *, 'Only the first 100 values will be read.'
      end if
c
      do 100, COUNT = 1, header
         read *, value
         X(COUNT) = value
  100 continue
c
      COUNT = header
c
      end
```

C.12. ALGORITHM 8–20 SUM THE VALUES IN AN ARRAY

In this subroutine, which sums the values in the array it is given as a precondition, the element type of array X could be either integer or real, as needed. We have chosen to use integer. The sum is declared to be of the same type as the element of the array because the sum of integers is an integer (likewise for reals). Thus the type of the sum should be the same as the type of the elements of the array.

Listing of the subroutine

```
      subroutine SumLst (X, N, sum)
c
      integer i, N
      integer sum, X(*)
c
      sum = 0
c
      do 100, i = 1, N
         sum = sum + X(i)
```

```
  100 continue
c
      end
```

C.13. ALGORITHM 8–21 PROGRAM TO AVERAGE A LIST OF VALUES

This program averages the values in a sentinel input list. (The algorithm referred to in the section heading is really a subalgorithm to average an array, but permits us to show a program with multiple subroutines.) We use the subroutines in previous sections to read a sentinel input list into an array and to sum the values in an array. The subalgorithm to average an array is given here as a subroutine as well. The main program calls upon the needed subroutines and does some error checking, as well as writes out the input list and the result.

Listing of the program

```
      program AverLs
c
      parameter (MAXIDX = 100)
      integer i, Length
      integer Values(MAXIDX)
      real Avg
c
      call RdSent (Values, Length)
      if (Length .le. 0) then
         print *, 'The data list is empty.'
      else
         print *, 'The input list is:'
         print *,
         do 100, i = 1, Length
            print *, Values(i)
  100    continue
         print *,
         call AvgLst (Values, Length, Avg)
         print *, 'The average of these values is: ', Avg
      end if
c
      end
c
c
c
      subroutine SumLst (X, N, sum)
c
```

```
      integer i, N
      integer sum, X(*)
c
      sum = 0
c
      do 100, i = 1, N
         sum = sum + X(i)
  100 continue
c
      end
c
c
c
      subroutine RdSent (X, COUNT)
c
      parameter (MAXIDX = 100)
      parameter (SENTNL = -1)
      integer COUNT
      integer value
      integer X(*)
c
      COUNT = 0
      read *, value
c
  100 if ((COUNT .lt. MAXIDX) .and. (value .ne. SENTNL)) then
         COUNT = COUNT + 1
         X(COUNT) = value
         read *, value
         goto 100
      end if
c
      if (value .ne. SENTNL) then
         print *, 'The data list is too long for this program.
         print *, 'Found by: read a sentinel list.'
      end if
c
      end
c
c
c
      subroutine AvgLst (LIST, N, AVERAG)
c
      integer N, TOTAL, LIST(*)
      real AVERAG
c
      if (N .gt. 0) then
         call SumLst (LIST, N, TOTAL)
```

```
            AVERAG = TOTAL/N
         else
            AVERAG = 0.0
            print *, 'An attempt to average an empty list'
            print *, 'was detected in AvgLst.'
         end if
c
         end
```

Listing of a sample data file

```
               76
               24
              -15
               27
                5
                0
               16
               21
                8
               -1
```

Listing of the output generated

```
The input list is:

                76
                24
               -15
                27
                 5
                 0
                16
                21
                 8

The average of these values is:        18.0000000
```

C.14. ALGORITHM 9–5 SEARCH A LIST FOR A PARTICULAR VALUE

This subroutine implements the search algorithm. The element type of the array may be any type whose values can be compared for equality by using the ".eq." relational operator. Y is declared to be of the same type as the array element type because it (and all the values in the array) must be of the same type. Y, after all, is to be compared with elements in X.

Listing of the subroutine

```
      subroutine Search (X, Place, Y, found, pos)
c
      integer Place, pos, scan
      integer Y, X(*)
      logical found
c
      found = .false.
      scan = 0
c
  100 if ( (scan .lt. Place) .and. (.not. found) ) then
         scan = scan + 1
         if ( X(scan) .eq. Y ) then
            found = .true.
         end if
         goto 100
      end if
c
      pos = scan
c
      end
```

C.15. ALGORITHM 9–9 USING PARALLEL ARRAYS, PRINT FREQUENCIES

This program constructs a list of values from the input list in such a way that only one instance of an input value occurs in an array. It also counts the number of times each value occurs. Described in the discussion of the algorithm as building a list of unique values and frequencies, this implementation uses parallel arrays.

Rather than a simple subroutine (the natural translation of the subalgorithm referred to in the section heading), we present a program that uses such a subroutine. We also use the search subroutine that was translated in the last section.

In this problem, we need two distinct arrays: one to hold the individual values and another to hold the count of the occurrences of the corresponding values. We have chosen the integer type for both.

Listing of the program

```
      program PFreq
c
      parameter (MAXIDX = 100)
      integer Length, i
      integer UnqVal(MAXIDX)
      integer Freqs(MAXIDX)
c
      call ConUnq (UnqVal, Freqs, Length)
      print *, 'The unique values and their frequency'
      print *,
      print *, '             value     frequency'
      print *,
c
      do 100, i = 1, Length
         print *, UnqVal(i), Freqs(i)
  100 continue
c
      end
c
c
c
      subroutine Search (X, Place, Y, found, pos)
c
      integer Place, pos, scan
      integer Y, X(*)
      logical found
c
      found = .false.
      scan = 0
c
  100 if ( (scan .lt. Place) .and. (.not. found) ) then
          scan = scan + 1
          if ( X(scan) .eq. Y ) then
             found = .true.
          end if
          goto 100
      end if
c
      pos = scan
c
      end
c
```

```
c
c
      subroutine ConUnq (unique, freq, size)
c
      integer size, pos
      integer unique(*), value
      integer freq(*)
      logical found
c
      size = 0
c
  100 if (size .lt. 100) then
        read (*, *, end=110) value
        call Search (unique, size, value, found, pos)
        if (.not. found) then
           size = size + 1
           unique(size) = value
           freq(size) = 1
         else
            freq(pos) = freq(pos) + 1
         end if
         goto 100
      end if
c
      print *, 'The data list is too long for this program.'
      print *, 'This error was found by ConUnq.'
c
  110 continue
c
      end
```

Listing of a sample data file

```
21
17
21
16
17
21
10
16
17
10
 5
 5
```

```
                                    10
                                     5
                                    17
                                    17
                                     8
                                    16
                                   100
                                     5
                                     8
```

Listing of the output generated

```
The unique values in the data and their frequency

              value    frequency
               21           3
               17           5
               16           3
               10           3
                5           4
                8           2
              100           1
```

C.16. ALGORITHM 9–13 PROGRAM TO SORT AN ARRAY

This program reads in an end-of-file input list and sorts it into ascending sequence. The subalgorithm referred to in the section heading (the insertion sort subalgorithm) has been translated into the subroutine Sort and includes the required subroutines to find the smallest element in a portion of the array, Small, and Switch, along with RdEOF. We wrote a main program to call the reading and sorting subroutines and to write out the resulting sorted array.

Listing of the program

```
      program SortLs
c
      parameter (MAXIDX = 100)
      integer DatLst(MAXIDX)
      integer Length, i
c
```

```
      call RdEOF (DatLst, Length)
      call Sort (DatLst, Length)
      print *, 'The sorted data list is:'
      print *,
c
      do 100, i = 1, Length
         print *, DatLst(i)
  100 continue
c
      end
c
c
c
      subroutine RdEOF (X, COUNT)
c
      parameter (MAXIDX = 100)
      integer COUNT
      integer value
      integer X(*)
c
      COUNT = 0
c
  100 if (COUNT .lt. MAXIDX) then
          read (*, *, end=110) value
          COUNT = COUNT + 1
          X(COUNT) = value
          goto 100
      end if
c
      print *, 'The data list is too long for this program.'
      print *, 'This error was found by: read an eof list.'
c
  110 continue
c
      end
c
c
c
      subroutine Switch (X, Y)
c
      integer X, Y, temp
c
      temp = X
      X = Y
      Y = temp
c
      end
```

```
c
c
c
      subroutine Small (X, N, K, p)
c
      integer N, K, p, i
      integer X(*)
c
      p = K
c
      do 100, i = K, N
         if (X(i) .lt. X(p)) then
            p = i
         end if
  100 continue
c
      end
c
c
c
      subroutine Sort (X, size)
c
      integer size, j, p
      integer X(*)
c
      do 100, j = 1, size-1
         call Small (X, size, j, p)
         call switch (X(j), X(p))
  100 continue
c
      end
```

Listing of a sample data file

```
          122
           76
           50
           81
           10
           36
            5
            0
          110
           75
           81
            6
           43
```

Listing of the output generated

```
The sorted data list is:
                    0
                    5
                    6
                   10
                   36
                   43
                   50
                   75
                   76
                   81
                   81
                  110
                  122
```

Bibliography

Adams, James L. *Conceptual Blockbusting*. New York: W. W. Norton, 1979.

Aho, A. V., J. E. Hopcraft, and J. D. Ullman. *The Design and Analysis of Computer Algorithms*. Reading, MA: Addison-Wesley, 1974.

Bailey, T. Eugene. *Program Design with Pseudocode*. Stillwater, OK: Kendall/Hunt, 1981.

Berztiss, A. T. *Data Structures Theory and Practice*. New York: Academic Press, 1971.

Brooks, Frederick P. *The Mythical Man-Month*. Reading, MA: Addison-Wesley, 1975.

Descartes, R. *OEuvres des Descartes*. Edited by Charles Adams and Paul Tannery. Paris: Leopold Cerf, 1906. (Vol. X contains the original, "Regulae ad Directionem Ingenii," referred to here as "Rules for the Direction of the Mind.")

Dromey, R. G. *How to Solve it by Computer*. Englewood Cliffs, NJ: Prentice-Hall International, 1982.

Ernst, George W., and Allen Newell. *GPS: A Case Study in Generality and Problem Solving*. New York: Academic Press, 1969.

Gardiner, Martin. *Aha! Insight*. New York: W. H. Freeman, 1978.

Gardiner, Martin. *Aha! Gotcha: Paradoxes to Puzzle and Delight*. San Francisco: W. H. Freeman, 1982.

Goldstein, H. H. *The Computer from Pascal to von Neumann*. Princeton, N.J.: Princeton University Press, 1972.

Grogono, Peter, and Sharon H. Nelson. *Problem Solving & Computer Programming*. Reading, MA: Addison-Wesley, 1982.

Gross, Jonathan, and Walter Brainerd. *Fundamental Programming Concepts*. New York: Harper and Row, 1972.

Horowitz, F., and S. Sahni. *Fundamentals of Data Structures*. Potomac, MD: Computer Science Press, 1976.

Kernighan, B. W., and P. J. Plauger. *The Elements of Programming Style*. New York: McGraw Hill, 1974. (Copyright by Bell Telephone Laboratories.)

Kernighan, B. W., and P. J. Plauger. *Software Tools*, Reading, MA: Addison-Wesley, 1976. (Copyright by Bell Telephone Laboratories and Yourdon Inc.)

Knuth, D. E. *The Art of Computer Programming*, 2nd ed. Reading, MA: Addison-Wesley, 1973.

Knuth, D. E. *Sorting and Searching*. Reading, MA: Addison-Wesley, 1973.

Leibnitz, W. *Philosophische Schriften*. Edited by Gerhardt. Berlin: Weidmannsche Buchhandlung, 1880.

Lewis, William. *Problem Solving for Programmers*. Rochelle Park, NJ: Hayden, 1980.

Mitchell, William. *Prelude to Programming*. Reston, VA: Reston Publishing, 1984.

Motil, John. *Programming Principles: an Introduction*. Boston, MA: Allyn and Bacon, 1984.

Newton, Sir Issac, *Universal Arithmetick*. Translated by Ralphson. London: W. Johnson, Ludgate-Street, 1769.

Polya, G. "Wie sucht Man die Loesung mathematischer Aufgaben?" *Zeitschrift fuer mathematischen und naturwissenschaftlichen Unterricht*, vol. 63 (1932), pp. 159-169.

Polya, G. "Let us Learn Guessing," In *Etudes de Philosophie des Sciences, en hommage à Ferdinand Gonseth*, pp. 147-154. Neuchatel, Switzerland: Editions du Griffon, 1950.

Polya, G., *Induction and Analogy in Mathematics*. Mathematics and Plausible Reasoning, vol. 1. Princeton, NJ: Princeton University Press, 1954.

Polya, G. *Patterns of Plausible Inference*, Mathematics and Plausible Reasoning, vol. 2. Princeton, NJ: Princeton University Press, 1954.

Polya, G. *How to Solve It*. 2nd ed. Princeton, NJ: Princeton University Press, 1957.

Polya, G. *Mathematical Discovery*. New York: John Wiley & Sons, 1981.

Schneider, G. M., and S. C. Bruell. *Advanced Programming and Problem Solving with Pascal*. New York: John Wiley & Sons, 1981.

Rubinstein, Moshe F. *Patterns of Problem Solving*. Los Angeles: Prentice-Hall, 1975.

Simon, Herbert. *The Sciences of the Artificial*. Cambridge, MA: M. I. T. Press, 1969.

Volkman, Arthur G., ed. *Thoreau On Man & Nature*. Mount Vernon, VA: Peter Pauper Press, 1960.

Watkins, R. P. *Computer Problem Solving*. Huntington, NY: Robert E. Krieger Publishing, 1980. (Copyright by John Wiley & Sons, 1974.)

Weinberg, G. M. *The Psychology of Computer Programming*. New York: Van Nostrand Reinhold, 1971.

Wickelgren, W. A. *How to Solve Problems: Elements of a Theory of Problems and Problem Solving*. San Francisco: W. H. Freeman, 1974.

Index